Aging of U.S. Air Force Aircraft

FINAL REPORT

Committee on Aging of U.S. Air Force Aircraft
National Materials Advisory Board
Commission on Engineering and Technical Systems
National Research Council

Publication NMAB-488-2
NATIONAL ACADEMY PRESS
Washington, D.C. 1997

NOTICE: The project that is the subject of this report was approved by the Governing Board of the National Research Council, whose members are drawn from the councils of the National Academy of Sciences, the National Academy of Engineering, and the Institute of Medicine. The members of the committee responsible for the report were chosen for their special competences and with regard for appropriate balance.

This report has been reviewed by a group other than the authors according to procedures approved by a Report Review Committee consisting of members of the National Academy of Sciences, the National Academy of Engineering, and the Institute of Medicine.

The National Academy of Sciences is a private, nonprofit, self-perpetuating society of distinguished scholars engaged in scientific and engineering research, dedicated to the furtherance of science and technology and to their use for the general welfare. Upon the authority of the charter granted to it by the Congress in 1863, the Academy has a mandate that requires it to advise the federal government on scientific and technical matters. Dr. Bruce M. Alberts is president of the National Academy of Sciences.

The National Academy of Engineering was established in 1964, under the charter of the National Academy of Sciences, as a parallel organization of outstanding engineers. It is autonomous in its administration and in the selection of its members, sharing with the National Academy of Sciences the responsibility for advising the federal government. The National Academy of Engineering also sponsors engineering programs aimed at meeting national needs, encourages education and research, and recognizes the superior achievements of engineers. Dr. William A. Wulf is interim president of the National Academy of Engineering.

The Institute of Medicine was established in 1970 by the National Academy of Sciences to secure the services of eminent members of appropriate professions in the examination of policy matters pertaining to the health of the public. The Institute acts under the responsibility given to the National Academy of Sciences by its congressional charter to be an adviser to the federal government and, upon its own initiative, to identify issues of medical care, research, and education. Dr. Kenneth I. Shine is president of the Institute of Medicine.

The National Research Council was organized by the National Academy of Sciences in 1916 to associate the broad community of science and technology with the Academy's purposes of furthering knowledge and advising the federal government. Functioning in accordance with general policies determined by the Academy, the Council has become the principal operating agency of both the National Academy of Sciences and the National Academy of Engineering in providing services to the government, the public, and the scientific and engineering communities. The Council is administered jointly by both Academies and the Institute of Medicine. Dr. Bruce M. Alberts and Dr. William A. Wulf are chairman and vice chairman, respectively, of the National Research Council.

This study by the National Materials Advisory Board was conducted under Contract No. F49620-96-C-0040 with the U.S. Air Force Office of Scientific Research. Any opinions, findings, conclusions, or recommendations expressed in this publication are those of the author(s) and do not necessarily reflect the views of the organizations or agencies that provided support for the project.

International Standard Book Number 0-309-05935-6

Available in limited supply from:
National Materials Advisory Board
2101 Constitution Avenue, N.W.
HA-262
Washington, DC 20418
202-334-3505

Additional copies are available for sale from:
National Academy Press
2101 Constitution Avenue, N.W.
Box 285
Washington, DC 20055
800-624-6242 or 202-334-3313 (in the Washington metropolitan area)

Copyright 1997 by the National Academy of Sciences. All rights reserved.

Cover: F-15 air superior fighters. U.S. Air Force photograph.

Printed in the United States of America.

COMMITTEE ON AGING OF U.S. AIR FORCE AIRCRAFT

CHARLES F. TIFFANY (chair), NAE, Boeing Military Airplanes (retired), Tucson, Arizona
SATYA N. ATLURI, NAE, Georgia Institute of Technology, Atlanta
CATHERINE A. BIGELOW, Federal Aviation Administration Technical Center, Atlantic City, New Jersey
EARL W. BRIESCH, Dayton Aerospace Inc., Dayton, Ohio
ROBERT J. BUCCI, Alcoa Technical Center, Alcoa Center, Pennsylvania
WENDY R. CIESLAK, Sandia National Laboratories, Albuquerque, New Mexico
EUGENE E. COVERT, NAE, Massachusetts Institute of Technology, Cambridge
B. BORO DJORDJEVIC, Johns Hopkins University, Baltimore, Maryland
CHARLES E. HARRIS, NASA Langley Research Center, Hampton, Virginia
JAMES W. MAR, NAE, Massachusetts Institute of Technology (retired), Pacific Grove, California
J. ARTHUR MARCEAU, Boeing Commercial Airplane Group, Seattle, Washington
CHARLES SAFF, Boeing Information, Space, and Defense Systems Group, St. Louis, Missouri
EDGAR A. STARKE, JR., University of Virginia, Charlottesville
DONALD O. THOMPSON, NAE, Iowa State University, Ames

National Materials Advisory Board Staff

THOMAS E. MUNNS, Senior Program Officer
AIDA C. NEEL, Senior Project Assistant
BONNIE SCARBOROUGH, Research Associate

National Materials Advisory Board Liaison

JAN D. ACHENBACH, NAS/NAE, Northwestern University, Evanston, Illinois

Air Force Science and Technology Board Liaison

ALTON D. ROMIG, JR., Sandia National Laboratories, Albuquerque, New Mexico

Air Force Technical Liaison

JIM C.I. CHANG, Office of Scientific Research, Washington, D.C.
WILLIAM R. ELLIOTT, Warner-Robins Air Logistics Center, Robins AFB, Georgia
JOSEPH P. GALLAGHER, Wright Laboratories, Wright-Patterson AFB, Ohio
JOHN W. LINCOLN, Aeronautical Systems Center, Wright-Patterson AFB, Ohio
DONALD PAUL, Wright Laboratories, Flight Dynamics Directorate, Wright-Patterson AFB, Ohio
VINCENT J. RUSSO, Wright Laboratories, Materials Directorate, Wright-Patterson AFB, Ohio
O. LESTER SMITHERS, Aeronautical Systems Center, Wright-Patterson AFB, Ohio

NATIONAL MATERIALS ADVISORY BOARD

ROBERT A. LAUDISE (chair), NAS/NAE, Bell Laboratories, Lucent Technologies, Murray Hill, New Jersey
G.J. ABBASCHIAN, University of Florida, Gainesville
JAN D. ACHENBACH, NAS/NAE, Northwestern University, Evanston, Illinois
MICHAEL I. BASKES, Sandia/Livermore National Laboratories, Livermore, California
JESSE L. BEAUCHAMP, NAS, California Institute of Technology, Pasadena
EDWARD C. DOWLING, Cyprus Amax Minerals Company, Englewood, Colorado
FRANCIS DISALVO, NAS, Cornell University, Ithaca, New York
ANTHONY G. EVANS, NAE, Harvard University, Cambridge, Massachusetts
JOHN A.S. GREEN, The Aluminum Association, Washington, D.C.
JOHN H. HOPPS, Morehouse College, Atlanta, Georgia
MICHAEL JAFFE, Hoechst Celanese Corporation, Summit, New Jersey
SYLVIA M. JOHNSON, SRI International, Menlo Park, California
LIONEL C. KIMERLING, Massachusetts Institute of Technology, Cambridge
HARRY A. LIPSITT, Wright State University, Dayton, Ohio
RICHARD S. MULLER, NAE, University of California, Berkeley
ELSA REICHMANIS, NAE, Bell Laboratories, Lucent Technologies, Murray Hill, New Jersey
KENNETH L. REIFSNIDER, Virginia Polytechnic Institute and State University, Blacksburg
EDGAR A. STARKE, JR., University of Virginia, Charlottesville
KATHLEEN C. TAYLOR, NAE, General Motors Corporation, Warren, Michigan
JAMES WAGNER, Johns Hopkins University, Baltimore, Maryland
JOSEPH WIRTH, Raychem Corporation, Menlo Park, California
BILL G.W. YEE, Pratt and Whitney, West Palm Beach, Florida

ROBERT E. SCHAFRIK, Director

Preface

The U.S. Air Force requested the National Research Council to identify research and development (R&D) needs and opportunities to support the continued operation of their aging aircraft. Specifically, this study focuses on aging aircraft structures and materials and has the major objectives of

1. developing an overall strategy that addresses the Air Force aging aircraft needs
2. recommending and prioritizing specific technology opportunities in the areas of

   - fatigue, corrosion fatigue, and stress corrosion cracking
   - corrosion prevention and mitigation
   - nondestructive inspection
   - maintenance and repair
   - failure analysis and life prediction methodologies

The approach that the committee took to accomplish this study was to conduct working sessions to identify current aging aircraft problems and technology needs; review ongoing and planned aging aircraft R&D efforts by the Air Force; and review related research at other government agencies, within industry, and in the academic research community.

The committee conducted a total of six meetings, prepared an interim report (NMAB-488-1), which was released in March 1997, and prepared this final report. In addition, numerous data-gathering discussions were held between individual committee members and various individuals from within the Air Force's research, engineering, logistics, and operational organizations. The purpose of the first meeting held at the Wright Aeronautical Laboratories, Wright-Patterson AFB, Ohio, was to review current and planned laboratory programs that are part of the Air Force aging aircraft program. The purpose of the second meeting, held at the San Antonio Air Logistics Center, Kelly AFB, Texas, was to identify the common problems associated with maintaining and operating aging systems and to review the applied R&D efforts under way at the Air Force air logistic centers (ALCs). Representatives from the five ALCs (i.e., Warner-Robins, Oklahoma City, San Antonio, Ogden, and Sacramento) participated in the meeting. At the third committee meeting, held in Washington, D.C., the committee reviewed ongoing and recently completed basic research programs at the Air Force Office of Scientific Research and developed the preliminary findings for the interim report. The fourth meeting was held in Irvine, California, at which time the committee reviewed related research being conducted by the National Aeronautics and Space Administration and the Federal Aviation Administration, finalized the interim report, and began developing recommendations for an overall aging aircraft strategy and identifying future research opportunities. At the fifth meeting, which was held in Washington, D.C., the committee reviewed related research being conducted by the Navy and received briefings on the F-15 aircraft structural history and on the aging of advanced composite structures. In addition, the committee continued their discussions on recommended strategy, research opportunities, and an approach for the prioritization of these opportunities. The sixth and final committee meeting was held in Washington, D.C., for the purpose of finalizing the prioritization of research opportunities and reviewing the initial draft of this final report.

The interim report that was released in March 1997 was prepared at the request of the Air Force research community and included the committee's preliminary technical assessment of the Air Force current aging aircraft R&D program. The report provided a description of the Air Force's aging aircraft problem from the force management perspective, a preliminary assessment of the force management process and its needs, a discussion of the key technical issues and apparent R&D needs, and a preliminary assessment of the current aging aircraft R&D program along with suggested areas of improvement and changes in emphasis.

As was pointed out in the preface to the committee's interim report, it became apparent very early in this study that the overall strategy to address the Air Force's aging aircraft needs must encompass much more than R&D needs and opportunities. There are a number of overarching engineering and management issues that also need to be addressed. These include issues involving the force management process, the continued enforcement of the Air Force's Aircraft Structural Integrity Program and its supporting structures and materials specifications, the need to update the durability and damage tolerance assessments of the aging aircraft, the need for increased emphasis on identifying and applying existing technologies to the Air Force's aging aircraft problems, the need for stable funding for technology transition at the Air Force's ALCs, and the technical skills needed to support the aging aircraft program. This final report presents an overall strategy that the committee believes addresses these issues as well as the near-term and long-term research and development needs and opportunities.

Charles F. Tiffany, Chair
Committee on Aging of U.S. Air Force Aircraft

Contents

EXECUTIVE SUMMARY 1

PART I PROBLEM DEFINITION AND STATUS OF THE AGING FORCE 9

  1 INTRODUCTION 11
      Background, 11
      Study Objectives, 11
  2 AGING AIRCRAFT PROBLEM 13
      Managing the Force Structure, 13
      Future Force Projections, 16
      Assessment of the Force Structural Management Process, 19
  3 CURRENT STRUCTURAL STATUS OF THE AGING FORCE 22
      Air Force-Supported Aircraft, 22
      Contractor Logistics-Supported Aircraft, 24
  4 TECHNICAL ISSUES AND OPERATIONAL NEEDS 27
      Corrosion, 27
      Stress Corrosion Cracking, 28
      Fatigue Cracking, 29
      Nondestructive Evaluation, 32
      Structural Maintenance and Repairs, 33

PART II RECOMMENDED STRATEGY AND OPPORTUNITIES FOR
        NEAR-TERM AND LONG-TERM RESEARCH 35

  5 ENGINEERING AND MANAGEMENT TASKS 39
      Update of Durability and Damage Tolerance Assessments, 39
      Update of Force Structural Maintenance Plans and Individual
        Aircraft Tracking Programs, 41
      Stress Corrosion Cracking Assessments, 42
      Improved Corrosion Control Programs, 43
      Economic Service Life Estimation, 45
      Continued Enforcement of the Aircraft Structural Integrity Program, 46
      Technical Oversight and Retention of Technical Capabilities, 47
      Technology Transition into Aging Aircraft, 48
  6 RESEARCH RECOMMENDATIONS: FATIGUE 49
      Low-Cycle Fatigue, 49
      High-Cycle Fatigue, 51
      Corrosion/Environmental Effects, 54
  7 RESEARCH RECOMMENDATIONS: CORROSION AND
      STRESS CORROSION CRACKING 57
      Corrosion Prevention and Control, 57
      Stress Corrosion Cracking, 60

8 RESEARCH RECOMMENDATIONS: NONDESTRUCTIVE EVALUATION
   AND MAINTENANCE TECHNOLOGY 63
   Nondestructive Evaluation, 63
   Maintenance and Repair, 68
9 PRIORITIZED RESEARCH RECOMMENDATIONS 73
   Critical Priorities, 73
   Near-Term Research, 73
   Long-Term Research, 74
10 FUTURE STRUCTURAL ISSUES: COMPOSITE PRIMARY
   STRUCTURES 76
   Applications and Service Experience, 76
   Recommendations for Long-Term Research, 77

REFERENCES 79

APPENDICES
   A SYNOPSES OF AIR FORCE AGING AIRCRAFT
     STRUCTURAL HISTORIES 87
   B BIOGRAPHICAL SKETCHES OF COMMITTEE MEMBERS 111

Tables and Figures

TABLES

ES-1 Priority-1 Near-Term and Long-Term Research Recommendations, 5
ES-2 Priority-2 Near-Term and Long-Term Research Recommendations, 6
ES-3 Priority-3 Near-Term and Long-Term Research Recommendations, 7

2-1 Tasks of the Air Force Aircraft Structural Integrity Program, 15

3-1 Data on Force Status for Air Force-Supported Aircraft, 23
3-2 Air Force Commercial-Derivative Aircraft Using Contractor Logistics Support, 24
3-3 Comparison between Utilization of Air Force CLS Aircraft and Commercial Equivalents, 25

5-1 Prioritization of DADTA Update Needs for Air Force-Supported Aircraft, 40

8-1 Critical NDE Inspection Needs for Aging Aircraft, 65

9-1 Prioritized Near-Term Research Recommendations, 74
9-2 Prioritized Long-Term Research Recommendations, 75

FIGURES

2-1 Force structure projection for the ACC fighter, bomber, and attack aircraft, 17
2-2 Force structure projection for the ACC airlift and rescue aircraft, 17
2-3 Force structure projection for other ACC aircraft, 18
2-4 Force structure projection for AMC aircraft, 18
2-5 Force structure projection for AFSOC aircraft, 19
2-6 Force structure projection for AETC aircraft, 20

II-1 Recommended overall strategy to address Air Force aging aircraft challenges, 36
II-2 Basic elements of the recommended near-term and long-term R&D programs, 37

5-1 Overall approach to durability and damage tolerance assessments, 40
5-2 Organization of commercial aircraft industry aging aircraft working groups, 44

A-1 C-5 flying hour distribution, 90
A-2 General locations for B-52G/H structural improvements, 92
A-3 B-52H current use rate, 93
A-4 F-15 buffet-induced problems, 96
A-5 F-16 structural arrangement, 97
A-6 F-16 structural modification areas, 98
A-7 A-10 structural arrangement, 99
A-8 Boeing 707 wing tear-down locations, 102
A-9 F-111 D6ac steel components, 105
A-10 Original lower wing skin design for the T-38 aircraft, 109

Acronyms

AATSG	aging aircraft technical steering group
AAWG	Airworthiness Assurance Working Group
ACC	Air Combat Command
ACI	analytical condition inspection
AETC	Air Education and Training Command
AFMPP	Air Force modernization planning process
AFR	Air Force regulation
AFSOC	Air Force Special Operations Command
ALC	air logistics center
AMC	Air Mobility Command
ASIP	Aircraft Structural Integrity Program
CACRC	Commercial Aircraft Composite Repair Committee
CLS	contractor logistics support
CPC	corrosion-preventive compound
CPCP	corrosion prevention and control program
DADTA	durability and damage tolerance assessment
EIF	equivalent initial flaw
FAA	Federal Aviation Administration
FAR	Federal Air Regulation
FSMP	force structural maintenance plan
IATP	individual aircraft tracking program
JPATS	Joint Primary Aircraft Training System
JSF	Joint Strike Fighter
LESS	loads/environment spectra study
LIF	lead-in fighter
NASA	National Aeronautics and Space Administration
NDE	nondestructive evaluation
NDI	nondestructive inspection
PDM	programmed depot maintenance
POD	probability of detection
R&D	research and development

SAB	Scientific Advisory Board
SCC	stress corrosion cracking
TIE	Technology and Industrial Support Engineering (ALC)
TPIPT	technology planning integrated product team
VOC	volatile organic compound
WFD	widespread fatigue damage

Executive Summary

The U.S. Air Force has many old (20 to 35+ years) aircraft that are the backbone of the total operational force, some of which will be retired and replaced with new aircraft. However, for the most part, replacements are a number of years away. For many aircraft, no replacements are planned, and many are expected to remain in service another 25 years or more.

To varying degrees, all of these older aircraft have encountered, or can be expected to encounter, aging problems such as fatigue cracking, stress corrosion cracking, corrosion, and wear. Through the Aircraft Structural Integrity Program (ASIP) and through durability and damage tolerance assessments (DADTAs) of older aircraft, the Air Force has already identified many potential problems, developed individual aircraft tracking programs, developed force structural maintenance plans, and taken maintenance actions to ensure safety and extend aircraft life. The Air Force has also initiated an aging aircraft research and development (R&D) program intended to support ASIP and address identified needs in the areas of widespread fatigue damage, corrosion–fatigue interactions, structural repairs, dynamics, health monitoring, nondestructive evaluation and inspection, and various aircraft subsystems.

The National Research Council Committee on Aging of U.S. Air Force Aircraft was formed to (1) identify Air Force aging aircraft needs and an overall strategy that addresses these needs and (2) recommend and prioritize specific technology opportunities, complementary to the efforts of industry, the Federal Aviation Administration, the National Aeronautics and Space Administration, and international organizations. The topics asked to be considered by the committee include fatigue, corrosion–fatigue interactions, and stress corrosion cracking; corrosion prevention and mitigation; nondestructive inspection; maintenance and repair; and failure analysis and life prediction technologies.

This report provides the committee's findings, including (1) a description and assessment of the Air Force aging aircraft problem and the force management process, (2) a detailed summary of the structural status of the aging force,[1] (3) a discussion of key technical issues and R&D needs, (4) a recommended overall strategy to address the Air Force aging aircraft problem, (5) recommendations for near-term engineering and management actions, and (6) prioritized near-term and long-term research recommendations. The committee's primary focus was on the deterioration of the metallic alloys used in the Air Force aging aircraft. Emerging issues concerning polymeric composite primary structures used in the Air Force's newer aircraft (e.g., the B-2 and F-22) are discussed in the final chapter of the report.

CONCLUSIONS

The challenge to the Air Force management and technical community is to meet the following objectives related to aging aircraft:

Objective A. Identify and correct structural deterioration that could threaten aircraft safety.

Objective B. Prevent or minimize structural deterioration that could become an excessive economic burden or could adversely affect force readiness.

Objective C. Predict, for the purpose of future force planning, when the maintenance burden will become so high, or the aircraft availability so poor, that it will no longer be viable to retain the aircraft in the inventory.

Safety

The structural safety of the Air Force's aircraft is vitally dependent on damage tolerance requirements that have been imposed through military standards and specifications as part of ASIP. These requirements allow the designer to use either of the following two design approaches:

- *Fail-safe design.* This approach, which relies on multiple, redundant load paths or crack arrest features, is used in commercial aircraft design and for most of the Air Force's large aircraft.
- *Safe crack growth design.* This approach has been used for much of the structure in high-performance combat aircraft where weight is a significant consideration. Engineering analysis must demonstrate that the maximum probable nondetectable initial manufacturing flaw will not grow to critical size (i.e., the size required

---

[1] Appendix A contains synopses of structural histories of Air Force-supported aircraft.

to cause failure) in any critical structural area during the operational life of the aircraft.

The committee concludes that, with increasing age and with changes in operation (or aircraft configuration) that increase the severity of the operational stress spectrum, the primary threats to structural safety arise from

- the onset of widespread fatigue damage (WFD) in fail-safe-designed structures
- the inexorable increase in the number of fatigue-critical areas in safe-crack-growth-designed structures and the potential for missing new areas as they develop

The primary technical needs for fail-safe designs are

- improved methods of predicting the onset of WFD in an accurate and timely manner. This involves the prediction of initiation and growth of small fatigue cracks (or the interpretation of full-scale fatigue test data and service fatigue data), the prediction of fail-safe residual strength, and the evaluation of the potential effects of environmentally induced corrosion on crack initiation and growth and residual strength.
- development and implementation of nondestructive evaluation (NDE) techniques that can rapidly detect small fatigue cracks over large areas of the structure prior to the onset of WFD. Methods to detect second- or inner-layer cracks and hidden corrosion that could lead to the initiation of cracks would be included.

The primary technical needs for safe crack growth structural designs are

- to identify the next most probable fatigue-critical areas in the structure through careful evaluation of past full-scale fatigue test results, service experience, service loading data (including dynamic loads), design details (including potential areas for hidden corrosion), and the results of stress analyses and strain surveys
- to perform simulative testing and crack growth analyses to establish safety limits and safety inspection requirements for all critical areas
- to investigate the potential effects of corrosion on those factors that could affect safety limits and safety inspection requirements
- to continue to improve methods of identifying fatigue-critical areas and flight load conditions to continue to improve NDE techniques that are sensitive enough to detect small cracks in multilayered and hidden structures to support safety inspections

Economics and Readiness

The economic burden associated with the inspection and repair of fatigue cracks can be expected to increase with age until the task of maintaining aircraft safety could become so overwhelming and the aircraft availability so poor that the continued operation of the aircraft is no longer viable. In addition, corrosion detection, repair, and component replacement can add significantly to or, in some cases, dominate the total structural maintenance burden.

The committee concludes that the major emphasis of the Air Force's technical and force management with regard to corrosion and stress corrosion cracking (SCC) should be focused on the early detection of corrosion and the implementation of effective corrosion control and mitigation practices so as to drastically reduce unscheduled repairs and replacement costs and aircraft downtime. Key technical issues and operational needs include

- the development of improved NDE techniques for the detection and rough quantification of hidden corrosion
- the classification of corrosion severity to provide guidance for maintenance
- the generalized application of corrosion-preventive compounds and the development of corrosion-preventive compounds that can be applied on external surfaces to protect unsealed joints and fasteners
- the development of a material and process substitution handbook and engineering guidelines for the replacement of components exhibiting corrosion and SCC with more-resistant materials and processes
- the development and application of materials and processes to inhibit SCC
- the development of technologies for the removal, surface preparation, and reapplication of surface finishes with improved corrosion-resistant finishes on existing aircraft
- the assessment of the potential use of the dehumidified storage of aircraft, where practical

The committee believes that fatigue cracking will occur eventually on all aging aircraft as flight hours increase. From an economic standpoint, the major impact for a fail-safe-designed structure occurs with the onset of WFD. For safe-crack-growth-designed structures, the major impact occurs when the structure exhibits a rapid increase in the number of fracture-critical areas. In both cases, a choice must be made to undertake major modifications, structural replacement, or retirement. Although it may not be possible to avoid reaching this point for any given aircraft, operational changes such as fuel management, gust avoidance, active or passive load alleviation systems, reduced pressurization, and flight restrictions to minimize flight in severe mission segments can reduce the rate of fatigue damage and delay expensive repair–replace–retire decisions. For aircraft that are approaching their economic service limit, these options should be considered to allow time for modification or replacement acquisition programs.

Force Management and Predicted Economic Service Life

The Air Force modernization planning process includes the essential elements for force structure planning and management, but, to be completely effective, it should significantly improve estimates of the probable economic service life of aging aircraft systems. There is no clear definition of all of the cost elements that contribute to the economic service life of an aircraft, nor is there a precise methodology for estimating when the costs of operating and maintaining a system will be high enough to warrant replacement. The committee believes that the development of an estimate of economic service life with metrics that integrate the effects of structural deterioration (i.e., from fatigue and corrosion) with economic considerations is essential to force management.

Future Structural Issues

Metallic alloy structures make up the vast majority of the airframes in the Air Force aging aircraft. However, more-recent aircraft have significant quantities of primary flight controls (C-17) and primary airframe structures (B-2, F-22) constructed from carbon-fiber-reinforced polymeric composites. Although limited Navy and commercial aircraft service experience with composite laminate primary structures has indicated very few occurrences of damage in primary structures, the Air Force needs to continue to monitor the performance of their composite components. Potential degradation mechanisms to monitor in the future for composite structural applications include (1) the development of transverse matrix cracking resulting from mechanical, thermal, or hygrothermal stresses; (2) the growth of impact damage under fatigue loading; (3) the growth of manufacturing-induced damage, especially from fastener installation; or (4) the development of corrosion in adjacent metal structures. The committee recommends that the Air Force undertake long-term research to monitor potential deterioration of composite structures, including the development of improved NDE methods, and to develop or improve maintenance and repair technologies, especially for composite primary structures.

RECOMMENDATIONS

The committee recommends that the Air Force adopt a three-pronged strategy that includes (1) near-term engineering and management tasks, (2) a near-term R&D program, and (3) a long-term R&D program. Engineering and management tasks are near-term actions (within three to five years) to improve the maintenance and force management of aging aircraft. Supporting the near-term engineering and management tasks are the near-term R&D efforts that the committee believes should be performed under the direction of Air Force laboratories or by supporting contractors and academic institutions. The long-term R&D program includes those efforts that the committee believes will take longer than three to five years to achieve a mature technology that could be adopted by industry or the Air Force air maintenance organizations but nevertheless should be initiated now or continued if they already have been initiated.

Near-Term Engineering and Management Tasks

The Air Force postproduction force management process, involving the implementation of inspections and modifications derived from the ASIP tasks and the results of DADTAs, has been a huge success in protecting the structural safety of the force aircraft for more than two decades. However, the committee is concerned that the extended use of old aircraft, coupled with the potentially adverse effects of reduced military budgets; reduced manpower; grade structure limitations; increased reliance on contractor maintenance; the elimination or relaxation of military regulations, standards, and specifications; and possible complacency of Air Force management, may make this past success rather fragile. The committee believes that it will take aggressive Air Force management and engineering actions to counter this deterioration in capability and loss in ASIP oversight and to prevent further deterioration in the future. The Air Force should continue to enforce ASIP and maintain sufficient resources to track the force aircraft, to keep DADTAs up to date, and to keep corrosion and stress corrosion cracking from becoming a structural safety issue. Also, sufficient resources should be maintained in R&D to support and improve the aging aircraft engineering, inspection, and maintenance and repair activities.

The committee identified the following engineering and management tasks. With the exception of the technology transition task, which is considered to be a continuous effort throughout the life of a weapon system, all of the near-term engineering and management tasks should be accomplished within five years.

- *Update of durability and damage tolerance assessments (DADTAs).* The committee recommends that the DADTAs of Air Force aircraft be periodically updated. In general, an update about every five years is appropriate. For Air Force-supported aircraft, the aircraft that should be given highest priority for DADTA updates include the A-10, F-16, U-2,[2] and T-38. The contractor logistics-supported, large commercial derivative aircraft that should be given highest priority for structural review are the high-use C-18, C-22, and possibly the VC-137. In addition, the committee recommends that damage tolerance surveys be conducted for utility and

---

[2] The U-2 was developed for the government and is logistically supported by Lockheed-Martin.

commuter class commercial-derivative aircraft to determine the need for DADTAs.
- *Update of force structural maintenance plans and individual aircraft tracking programs.* Following the completion of the updates of the DADTAs (1) the inspection and modification requirements in the force structural maintenance plans should be updated to reflect any changes in the baseline operational spectra and any additional critical areas that were identified, and (2) an individual aircraft tracking program for each aircraft weapon system should be updated to reflect additional critical areas that need to be tracked.
- *Stress corrosion cracking (SCC) assessments.* The committee recommends that the Air Force include an assessment of the vulnerability of each of their aging aircraft to structural failure caused by SCC or SCC combined with fatigue as part of the proposed DADTA updates. Specifically, it is suggested that (1) stress corrosion critical areas be identified based on past service experience, the susceptibility of the materials to SCC, grain orientations, and probable levels of both applied and residual stresses; (2) an evaluation be made of potential failure modes and consequences of failure for each stress corrosion critical area; and (3) protection, inspection, modification, and replacement alternatives be developed as necessary.
- *Improved corrosion prevention and control programs.* The committee recommends that the Air Force (1) perform an internal audit of each of the commercial-derivative aging aircraft to ensure that the corrosion control programs are in full compliance with the mandated programs for the commercial counterparts; (2) review the detailed corrosion control programs of each of the Air Force developed aging aircraft and upgrade them as necessary to a level equivalent or better than the mandated programs for commercial aircraft; and (3) evaluate the applicability and cost effectiveness of dehumidification to reduce the likelihood of corrosion.
- *Economic service life estimation.* The committee recommends that the Air Force make a concerted effort to develop a credible service life estimation methodology, analogous to the cost and operational effectiveness analysis that is done early in a weapon system acquisition cycle, as the authoritative guide for supporting replacement decisions and budget inputs.
- *Continued enforcement of ASIP.* ASIP, as enforced through MIL-STD-1530 and supporting specifications, will no longer be placed on aircraft acquisition and modification contracts because of initiatives to reduce the use of government specifications in acquisition programs. The committee recommends that the Air Force take the lead in pursuing the development of a National Aerospace Standard to establish enforceable consensus industry standards for ASIP.
- *Technical oversight and retention of technical capabilities.* Reductions in technical capabilities and technical oversight should be addressed by (1) forming an aging aircraft engineering resources group to examine and develop solution options to engineering skill deficiencies (quantity and quality) in each of the aging aircraft disciplines, (2) forming an aging aircraft technical steering group to monitor and provide guidance to the various recommended near-term engineering and near- and long-term research activities discussed in this report, (3) forming five technical working groups (i.e., one for each of the five topical areas in the proposed near-term and long-term R&D programs) to provide the technical link from basic research through implementation, and (4) appointing a single knowledgeable and experienced technical leader responsible for the oversight of the aging aircraft engineering and the near-term and long-term R&D activities.
- *Technology transition into aging aircraft.* The committee recommends that generic aging aircraft technology programs with potential for wide application not be approved through the Air Force technology master process unless there is a clear link to an appropriate technology implementation program. It is critical to the success of the aging aircraft program that a seamless funding–budgeting link be created from development through application.

Near-Term and Long-Term Research

The committee developed prioritized recommendations for near-term (to support near-term engineering actions in the next five years) and long-term (more than five years until implementation) R&D in five program areas:

- *Fatigue* (including low-cycle fatigue, high-cycle fatigue, and corrosion/environmental effects):

  – despite efforts by the committee to develop research initiatives to improve the current approach to identify new fatigue-critical areas (i.e., analysis of full-scale fatigue test data and service experience), no viable near-term or long-term R&D activities were identified. The committee emphasizes the extreme importance of using all available data and up-to-date stress analysis methods to accomplish this task during the recommended DADTA updates, particularly for safe-crack-growth- (i.e., non-fail-safe-) designed aircraft.
  – in the area of low-cycle fatigue of fail-safe designed aircraft, the committee recommends near-term R&D to assess the validity of (and if necessary, develop improvements to) the current approach to estimate the onset of WFD and longer-term R&D to

analytically predict the initiation and growth of cracks to the sizes at onset of WFD
- in the area of high-cycle fatigue, the committee recommends near-term R&D to improve methods to determine dynamic response and long-term research to characterize threshold crack growth behavior, develop an analytical prediction of dynamic response, develop expert systems for the design and analysis of repairs, and develop dynamic load monitoring and alleviation
- in the area of corrosion/environmental effects, the committee recommends near-term R&D to assess the effects of prior corrosion on the fatigue crack growth and fracture behavior of airframe structural components and long-term fundamental research to provide an understanding of corrosion degradation mechanisms

- *Corrosion prevention and control.* The committee recommends near-term program emphasis on corrosion detection and maintenance technology (i.e., how to deal with existing corrosion) and longer-term emphasis on the fundamental understanding of corrosion and characterization of corrosion rates and the development and institutionalization of corrosion prevention and control practices.

- *Stress corrosion cracking.* The committee recommends that near-term R&D focus on developing data and documenting results that would lead to affordable upgrades in SCC prevention and component repair and modification procedures. The recommended focus of the long-term R&D is on establishing fundamental materials and microstructural effects on SCC susceptibility and a basic understanding of SCC mechanisms to support efforts in SCC prevention.

- *Nondestructive evaluation.* The committee recommends that near-term R&D emphasize the implementation of advances from related government and industry programs and an evaluation of NDE reliability for current methods as they apply to aging aircraft. Long-term R&D would emphasize the development of new NDE equipment and the application of computational methods and simulations in the development and evaluation of inspection techniques.

- *Maintenance and repair.* The committee recommends that the primary focus of the near-term programs be to apply the lessons learned from recent programs (e.g., C-141 and battle damage repair) for use at Air Force maintenance organizations. The recommended long-term focus is on the development of analytical design, structural assessment, and life prediction tools for repairs and repaired structures and to develop improved

TABLE ES-1 Priority-1 Near-Term and Long-Term Research Recommendations

Recommendation	Description	Objective	Timing
Fatigue			
None			
Corrosion Prevention and Control			
Evaluate durability of new protective coatings	Page 58	B	Near term
Basic research in corrosion prevention and control	Page 59	B	Long term
Basic research in coating durability	Page 60	B	Long term
Stress Corrosion Cracking			
Affordable upgrades in SCC prevention	Page 60	B	Near term
Evaluation of SCC protection systems	Page 60	B	Near term
Fundamental research in SCC prevention	Page 61	B	Long term
Nondestructive Evaluation			
Evaluate, validate, and implement NDE equipment and methods	Page 64	B	Near term
Develop integrated quantitative NDE capability	Page 66	B	Long term
Automation of wide-area NDE inspections	Page 68	B	Long term
Maintenance and Repair			
None			

TABLE ES-2 Priority-2 Near-Term and Long-Term Research Recommendations

Recommendation	Description	Objective	Timing
Fatigue			
Fail-safe residual strength prediction methods	Page 50	A	Near term
Improve current methods to estimate onset of WFD	Page 50	A	Near term
Methods to predict dynamic responses	Page 52	B	Near term
Effect of joint pillowing on fail-safety	Page 55	A	Near term
WFD crack formation and distribution relationships	Page 50	A	Long term
Analytical prediction of WFD crack distribution functions	Page 51	A	Long term
Validation of analytical WFD methods	Page 51	A	Long term
Crack growth threshold behavior	Page 52	B	Long term
Analytical methods to predict dynamic behavior	Page 53	B	Long term
Dynamic load monitoring and alleviation	Page 53	B	Long term
Effect of environment on growth of small cracks	Page 55	A	Long term
Effect of flaw morphology on crack growth	Page 56	A	Long term
Corrosion Prevention and Control			
Laboratory test protocol for accelerated corrosion testing	Page 57	B	Near term
Methods for early detection of corrosion	Page 58	B	Near term
Corrosion rates for major corrosion types	Page 59	B	Long term
Stress Corrosion Cracking			
Residual stresses and their alleviation	Page 61	A	Near term
SCC susceptibility of Air Force alloys	Page 61	A	Near term
Life prediction methods for SCC	Page 62	B	Long term
Nondestructive Evaluation			
NDE automation, data processing, and analysis	Page 66	B	Near term
Hybrid inspection technologies	Page 67	B	Long term
NDE to assess composite repairs	Page 67	B	Long term
Maintenance and Repair			
Guidelines to implement advances in bonded repairs	Page 69	B	Near term
Solid model interfaces to simulate repair methods	Page 70	B	Near term
Reduce cost of materials and structures substitution	Page 71	B	Near term
Repair design guidelines for high-cycle fatigue problems	Page 71	B	Near term
Expert system for design and analysis of repairs	Page 71	B	Long term
Common database for repair lessons learned	Page 72	B	Long term

damping materials for repair of structure prone to high-cycle fatigue.

Priority levels for recommended R&D opportunities were established relative to the Air Force objectives (i.e., safety of flight [Objective A] and maintenance costs and force readiness [Objective B]). Definitions of priority categories include

Critical priority: essential to flight safety (Objective A) (i.e., would eliminate a substantial threat to flight safety)

Priority 1: essential to the reduction of maintenance costs and improvement of force readiness (Objective B) (i.e., would enable the Air Force to address significant technical problems)

Priority 2: important to improved flight safety (Objective A) or reduced maintenance costs and improved force readiness (Objective B) (i.e., would represent significant improvement over current solutions)

Priority 3: advantageous to improved flight safety (Objective A) or reduced maintenance costs and improved force readiness (Objective B) (i.e., would improve efficiency or reduce cost of current methods)

There are no R&D efforts identified at this time that are of sufficient magnitude to be categorized as critical priority.

TABLE ES-3 Priority-3 Near-Term and Long-Term Research Recommendations

Recommendation	Description	Objective	Timing
Fatigue			
Effect of corrosion on material properties	Page 55	A	Near term
Effect of corrosion and corrosive environment on safety limits	Page 55	A	Near term
Expert systems for high-cycle fatigue repairs	Page 53	B	Long term
Effect of hydrogen on fatigue crack growth	Page 56	A	Long term
Corrosion Prevention and Control			
None			
Stress Corrosion Cracking			
None			
Nondestructive Evaluation			
Advanced technologies to track maintenance trends	Page 68	B	Long term
NDE for early corrosion detection	Page 68	B	Long term
Maintenance and Repair			
Guidelines on relative lives of bolted repairs	Page 70	A	Near term
Analysis methods for structural repairs	Page 72	B	Long term
Damping materials for dynamically loaded structure	Page 72	B	Long term

However, the committee believes that it is possible that the recommended DADTA updates, and in particular the high-priority updates on the A-10, F-16, U-2, and T-38, will identify critical-priority near-term R&D or engineering tasks. For example, these could involve the need to develop a specific inspection technique or a specific type of modification for one or more aircraft.

The committee's priority-1, -2, and -3 near-term and long-term R&D opportunities are summarized in Tables ES-1, ES-2, and ES-3, respectively. Each table contains reference to the pages in the report where the full description, background information, and justification for each recommendation can be found. As can be seen from the tables, there are a total of 9 priority-1 recommendations, 27 priority-2 recommendations, and 9 priority-3 recommendations. The priority-1 recommendations focus on reducing maintenance costs and improving force readiness (Objective B), particularly in the areas of corrosion prevention and control, SCC, and NDE. Many of the priority-2 recommendations address improving safety (Objective A) through development of improved methods to evaluate and analyze fatigue and stress corrosion cracking. The remainder of the recommendations deal with improvements in maintenance costs and force readiness (Objective B). Likewise, priority-3 recommendations address both objectives.

The 45 recommended research opportunities, when coupled with the 8 engineering and management tasks (which the committee considers to be essential), will substantially enhance the ability of the Air Force to address the aging aircraft problem and to sustain the forces well into the next century.

I

PROBLEM DEFINITION AND STATUS OF THE AGING FORCE

In Part I of the report, the committee provides the basis for this study of the U.S. Air Force aging aircraft. Chapter 1 describes the background and objectives for the study. Chapter 2 is a summary of force management processes, future force projections, and an assessment of force management, including the Aircraft Structural Integrity Program. Chapter 3 describes the current structural status of the aging aircraft in the Air Force inventory, including Air Force-supported aircraft and contractor logistics-supported aircraft. Finally, Chapter 4 identifies the technical issues and operational needs associated with Air Force aging aircraft that form the basis for the recommended engineering, management, and research and development tasks presented in Part II.

1
Introduction

BACKGROUND

The U.S. Air Force has many old aircraft that form the backbone of the total operational force structure. The oldest are the more than 500 jet tanker aircraft, the KC-135, that were first introduced into service more than 40 years ago. The B-52H bomber, the C-130 airlifter, the T-38 trainer, and the T-37A primary trainer all became operational 35 to 40 years ago; the C-141 and C-5A airlifters, 25 to 35 years ago; the F-15 air superiority fighter, the A-10 close air support aircraft, and the E-3 (AWACS), 20 to 25 years ago; and the F-16 multirole fighter and the KC-10 jet tanker, 15 to 20 years ago. Of these, only the C-141 is currently being replaced (by the C-17). Other replacements are in various stages of development for the T-37A (by the JPATS), the F-15 (by the F-22), and the F-16 (by the Joint Strike Fighter). For the most part, these replacements are a number of years away, and the program schedules continue to be constrained by and subject to the vagaries of annual funding cycles. For example, at best, it will be at least another 15 to 20 years before there will be a significant number of replacements for the F-16 combat fighter force. The remainder of the aircraft mentioned above have no planned replacements and are expected to remain in service an additional 25 years or more.

To varying degrees, all of these older aircraft either have encountered, or can be expected to encounter, aging problems such as fatigue cracking, stress corrosion cracking, corrosion, and wear. The challenge to the Air Force management and technical community is to meet the following objectives:

Objective A. Identify and correct problems that could threaten aircraft safety.

Objective B. Prevent or minimize problems that could become an excessive economic burden or adversely affect force readiness.

Objective C. For the purpose of future force planning, have the methodology to predict when the maintenance burden will become so high, or the aircraft availability so poor, that it will no longer be viable to retain the aircraft in the inventory.

The Air Force has been aware of these objectives for a number of years and has, through their Aircraft Structural Integrity Program (ASIP) and durability and damage tolerance assessments of their older aircraft, already identified many potential problems, developed individual aircraft tracking programs, developed force structural maintenance plans, and taken many maintenance actions to protect safety and extend aircraft life. The Air Force has also initiated an aging aircraft research and development (R&D) program that is intended to support ASIP and address identified needs in the areas of widespread fatigue damage, corrosion–fatigue interactions, structural repairs, dynamics, health monitoring, nondestructive evaluation and inspection (NDE/I), and various aircraft subsystems (Rudd, 1996).

STUDY OBJECTIVES

The Air Force requested that the National Research Council, through the National Materials Advisory Board, conduct this study with the following specific objectives:

- identify an overall strategy that addresses the Air Force aging aircraft needs
- recommend and prioritize specific technology opportunities in (1) fatigue, corrosion–fatigue interactions, and stress corrosion cracking; (2) corrosion prevention and mitigation; (3) nondestructive inspection; (4) maintenance and repair; and (5) failure analysis and life prediction technologies
- complement, rather than duplicate, the efforts of industry, the Federal Aviation Administration, the National Aeronautics and Space Administration, other military services, and international organizations

To address the overall objectives, the committee performed the following tasks:

- reviewed and analyzed the structural histories, problems, and force management procedures employed on the older Air Force aircraft to assist in identifying research needs and in developing a recommended overall aging aircraft strategy
- reviewed and analyzed critical degradation and failure mechanisms associated with aging aircraft that have been experienced to date or can be expected to be experienced in the future

- reviewed and evaluated corrosion prevention and mitigation procedures and methods that are applicable to aging aircraft, assessed ongoing research, and identified additional opportunities
- reviewed and evaluated methods for the inspection of structures for the purpose of assessing the adequacy of current methods and identified promising advanced NDE/I methods
- reviewed and evaluated repair methods and analysis procedures for the purpose of identifying possible deficiencies and promising research that could lead to their correction
- reviewed the current state of the art in failure analysis, life prediction, and structural risk assessment methodologies; identified deficiencies; assessed ongoing research; and identified additional research opportunities

This study emphasized aging of current airframe structures, especially aluminum primary structures. The important issues related to the aging of aircraft engine structures were not included in the committee's task and would be best addressed separately in the future.

This report summarizes the committee's assessment of the adequacy of the Air Force R&D program, assesses the force management process and its needs, identifies key technology issues and R&D needs, identifies and prioritizes R&D opportunities, and identifies and develops an overall strategy that addresses the Air Force's aging aircraft needs. The committee prepared a interim report (NRC, 1997) that focused on a preliminary assessment of the needs and an assessment of the current Air Force R&D program. This final report expands on this previous publication.

2

Aging Aircraft Problem

Any discussion of the wisdom of maintaining capital equipment is usually based on economic arguments. For example, if the cost of maintaining the equipment on a monthly or annualized basis exceeds the capital, interest, and amortization charges on replacement equipment, the decision to purchase the replacement is straightforward. Often the replacement equipment offers an improved productivity as well.

In the case of Air Force aircraft, safety-of-flight considerations also enter into the decision to repair or replace. Fortunately, inspection and maintenance procedures and the Aircraft Structural Integrity Program (ASIP) have been developed to reduce the likelihood of structural failure during the design service life. However, several external political changes, including the end of the Cold War, have caused the Air Force to change their approach to force management. As a result, the Air Force budget to develop new aircraft systems has been reduced. Because strategic policies have not been altered greatly, Air Force managers have concluded that the only way to meet the mission demands is to extend the service life of some of their aircraft forces.

Ultimately these factors will impact the force planning process. The Air Force aging aircraft problem can be best understood by examining the existing force management process, future force projections, and the current structural condition of the many types of aging aircraft in the Air Force inventory. In this chapter, force management processes and future force projections are summarized, followed by the committee's assessment of the process, including ASIP. A summary of the current structural condition of the many types of aging aircraft in the Air Force inventory is provided in Chapter 3. The committee's assessments of key technical issues related to the aging aircraft problem are discussed in Chapter 4.

MANAGING THE FORCE STRUCTURE

Modernization Planning Process

The Air Force modernization planning process (AFMPP) is the mechanism for supporting the five core competencies—air superiority, space superiority, precision employment, global mobility, and information dominance—provided by executive guidance documents. Aircraft systems are involved primarily in three of the five competencies: air superiority, precision employment, and global mobility.

The AFMPP integrates the elements that provide the foundation for the five competencies into a coherent modernization plan that reaches 25 years into the future. The foundation elements included in the modernization plan are

- readiness and sustainment
- research, development, test, and evaluation
- logistics
- information technology
- equipment and facilities
- manpower

The effectiveness of the aircraft systems (as well as other systems) in providing those competencies is determined largely by how well the foundation elements are integrated and addressed.

The key focus of the AFMPP is "modernization." Historically, the Air Force has been the world's technological leader in aircraft systems. This has been achieved through a robust science and technology program combined with frequent replacement of aging systems with new or modernized systems. This rapid replacement has slowed significantly in recent years because of budget constraints and affordability considerations. The result has been a shift to increased upgrading and life extension of many systems beyond what was typically done in the past.

The extended use of many aircraft results in increased maintenance and repair costs because of structural cracking and corrosion problems. In most cases, older aircraft spend longer times undergoing depot maintenance, with a resulting severe impact on readiness. Furthermore, extended aircraft service places increased importance on forecasting when the system must be replaced, either because of obsolescence or economic reasons (or a combination of both). If a system must be retired before the expected forecast service life, readiness could be impacted severely because a replacement system would not be ready in time to close the gap. Extended production lead times and budget exigencies for new systems make it even more important that the Air Force accurately determine, with a high degree of precision and confidence, the expected structural life of aircraft systems and the economics of sustaining them.

The AFMPP consists of the following six elements:

- *Mission area analysis.* Each of the major operating commands within the Air Force performs a mission area analysis that evaluates the military strategy provided by the chairman of the Joint Chiefs of Staff for new or changed missions. This review results in new or changed military tasks that the operating commands within the Air Force must then perform.
- *Mission needs analysis.* The operating commands evaluate their ability to accomplish assigned tasks (from the mission area analysis) and identify any issues. In performing this "task-to-need" analysis, the operating commands employ a variety of analytical tools. These analyses identify task deficiencies and possible nonmatériel or matériel solutions. If new or modified hardware is required, the mission need is documented in a mission needs statement.
- *Mission area plan.* The results of the mission area and mission needs analyses will be used to document, for the next 25 years, the most cost-effective means of correcting task deficiencies. The corrective actions could include nonmatériel solutions (e.g., changes to tactics and training), changes in force structure, system modification or upgrades, science and technology applications, or new hardware acquisition. Of interest are those solutions involving modifications, upgrades, new technology, and new hardware. Determining which solution is optimum, and planning for it, is greatly influenced by the aging aircraft problem.
- *Technology planning integrated product teams (TPIPTs).* TPIPTs provide the vital link to ensure that research efforts are responsive to user needs. In 1996 there were 21 different TPIPTs, each focused in a different mission area and involving all of the operating commands. The teams are administered by the Air Force Matériel Command's product centers. Eight of these TPIPTs are involved directly in aircraft system planning. Each team is responsible for coordinating the technology needs inputs among the technical and logistics communities and introducing them into the operating command's planning process. This is the primary mechanism for entering aging aircraft technology needs into the overall technology planning process. Each TPIPT documents identified needs in a development plan and issues a technology investment recommendation report. These reports serve as input into the technology master process.
- *Technology master process.* The technology master process is the vehicle through which technology strategy is planned and executed, based on the identified needs of the aircraft operators and system program directors. This process is designed to be comprehensive by including technology development, transition, application, and transfer. The process ensures that (1) all technologies are identified and prioritized for action, (2) budget-constrained technology projects are formulated in a highly integrated manner with full participation by the stakeholders, and (3) technologies that are candidates for application and insertion are validated and ready to enter the full acquisition cycle.
- *Aging Aircraft Office.* The Aging Aircraft Office is a recent addition to the Air Force planning process in support of aging aircraft. This office was established by the Air Force Matériel Command in April 1996 to address recognized problems in planning and executing technology programs in support of aging aircraft. The mission of the Aging Aircraft Office is to work within and outside the Air Force to implement technologies that extend the lives and reduce the cost of operating and maintaining aging aircraft systems. This office is intended to fill an important gap in the overall AFMPP by focusing attention on aging aircraft technologies and ensuring the expeditious implementation of needed technologies.

Service Life Projection during Acquisition

The expected service life for new aircraft is developed through the AFMPP, as was discussed in the preceding section. These goals are summarized in the systems requirements document that also describes the systems design parameters. The aircraft system operator then establishes the mission profiles for the new aircraft system. System design engineers use these mission profiles to develop load spectra, a critical input to the structural design activity. It is also during this time that a damage-tolerance-based ASIP, summarized in Table 2-1, is initiated.

The full-scale durability test task is especially significant with regard to the aging aircraft problem because this testing validates the design service life based on the operator's planned mission profiles. Full-scale durability test results establish the baseline from which the service life is updated throughout the service life cycle as mission profiles and use rates change. Data from the full-scale durability tests also assist in the development of the structural maintenance program required throughout the aircraft life cycle. The initial estimated weapon system phase-out point is also established based on estimates of safety limits and economic life considerations discussed in Chapter 4.

Few aircraft systems have been used as originally projected. This has resulted in the need for service life extension, modifications, or repair actions in advance of the originally projected time frame. The need for life extension, modification, or repair typically has been a result of corrosion or early fatigue damage caused by increased heavy use. Corrosion, unlike fatigue damage, has not been forecast analytically as

TABLE 2-1 Tasks of the Air Force Aircraft Structural Integrity Program

Task I: Design Information	Task II: Design Analysis and Development Tests	Task III: Full-Scale Testing	Task IV: Force Management Data Package	Task V: Force Management
ASIP master plan	Materials and joint allowables	Static	Final analyses	Loads and environment spectra survey
Structural design criteria	Loads analysis	Durability	Strength summary	Individual aircraft tracking data
Damage tolerance and durability control plans	Design service loads spectra	Damage tolerance	Force structural maintenance plan	Individual aircraft maintenance times
Selection of materials, processes, and joining methods	Design chemical/thermal environment spectra	Flight and ground operations	Loads and environment spectra survey	Structural maintenance records
Design service life and design use	Stress analysis	Sonic	Individual aircraft tracking program	
	Damage tolerance analysis	Flight vibration		
	Durability analysis	Flutter		
	Sonic analysis	Interpretation and evaluation of test results		
	Vibration analysis			
	Flutter analysis			
	Nuclear weapons effects analysis			
	Non-nuclear weapons effects analysis			
	Design development tests			

Source: Lincoln (1996).

part of ASIP. However, historically corrosion has caused an escalation of maintenance costs and, in many cases, severely impacted readiness because of the increased time required in depot repair. Some aircraft have been retired earlier than desired by the force planners because these aircraft became unaffordable because of escalating maintenance costs. Thus, the force management task of ASIP is tied closely to the aircraft aging problem.

Postproduction Force Management

During the deployment phase of the acquisition cycle, the Air Force establishes a maintenance and sustainment program. This program is based in part on data packages generated by ASIP. The maintenance and sustainment program includes such elements as

- field maintenance programs
- depot maintenance programs
- modification and repair programs
- technology transition and insertion
- weapon system assessments
- ASIP updating

It is the responsibility of the system program director to ensure that ASIP is continued on the weapon system throughout its operational life. This is done through the implementation of the force structural maintenance plan (FSMP), scheduling the required structural inspections and maintenance for the individual aircraft, maintaining structural maintenance records, conducting the loads/environment spectra survey (LESS), and implementing the individual aircraft tracking program (IATP). The system program director must ensure that durability and damage tolerance analyses are performed to provide new inspection and modification requirements if

- significant changes in use are noted from the LESS or the IATP
- new and unanticipated fatigue-critical areas show up in service aircraft

- there are significant structural configuration changes as a result of structural repairs or operational capability enhancement modifications

A durability and damage tolerance analysis may consist of an update of the original analysis conducted either during the design of the aircraft or during a durability and damage tolerance assessment (DADTA) performed subsequent to the initial design.[1]

In addition to the continuous enforcement of ASIP throughout the operational life of the aircraft and the translation of appropriate inspections to the field level, the system program director is also responsible for enforcing corrosion inspection and maintenance requirements and, as appropriate, translating them to the field level. The field-level ASIP and corrosion tasks can involve the use of standard or specialized nondestructive inspection (NDI) equipment and inspection criteria. The need for field-level repairs may result from these inspections. It is at this field level that insufficient NDI reliability of detection has been a problem. If the NDI techniques are overly complex or tedious and the failure data reporting burdensome, the resulting quality and reliability of both are dramatically reduced. This is also true, to some extent, for depot-level maintenance.

The system program director is also responsible for implementing the depot maintenance program on the weapon system. This typically consists of programmed depot maintenance (PDM) along with analytical condition inspection (ACI). The ACI tasks are performed annually on a selected small sample of the force. PDM covers the entire force, with a portion of the force scheduled for maintenance each year. In some cases, such as the newer systems, PDM may not exist. In these cases, the aircraft are usually sent to the depot for modifications and updates, along with an ACI. Speedlines and depot field teams are also employed to accelerate modifications. It is during these events that logistics engineers can gain significant insight into potentially life-limiting structural problems. As a result, the system program director can institute significant structural inspections, along with repair and modification programs, that add considerably to the cost of maintaining the weapon system. Therefore, improved technology relating to aging can have the greatest impact in the depot. Improvements in prediction of aging, detection of corrosion and fatigue cracks, and repair technologies can significantly reduce costs, extend service life, and enhance aircraft readiness.

As necessary, special modification and repair programs to accelerate the correction of structural cracking or corrosion are established to augment the PDM. The system program director creates a dedicated speedline at the depot or, in some cases, sends depot personnel to the operational bases as field teams. Speedlines and field teams often rely heavily on specialized NDI techniques tailored for the specific problems and perform specialized repairs when defects are discovered. For NDI and repair techniques, speed, ease of use, and reliability of detection become critical during speedline or field team activity because they directly affect the aircraft inspection and repair flow time and consequently the total time that the aircraft is out of service. Speedlines and field teams require additional aircraft to be taken out of service, impacting availability and readiness. Therefore, improving the technology for both NDI and maintenance and repair can result in significant benefits to aging aircraft systems.

Periodically, system program directors assess the health of aircraft over a broad range of indicators and report the status to senior management of the Air Force. One aspect of the assessment is the forecast life of the aircraft and problems affecting longevity. This process provides the system program director the opportunity to gain support to resolve aging aircraft problems and essential information to planners to ensure accurate force structure planning.

FUTURE FORCE PROJECTIONS

There are four major commands within the Air Force that operate aircraft systems to accomplish a wide range of missions. The Air Combat Command (ACC), Air Mobility Command (AMC), Air Force Special Operations Command (AFSOC), and Air Education and Training Command (AETC) all have long-range plans that include the phase-out of existing systems and the phase-in of new systems developed through the mission area planning process described above. Preliminary force structure plans can be summarized best through the use of "sand" charts that cover the next 25 years. These charts illustrate the life extension of existing systems and the introduction of relatively few new aircraft systems as replacements. The sand charts for each of the major operating commands, along with a brief description of each, are shown in the following sections.

Air Combat Command

Figures 2-1, 2-2, and 2-3 show the ACC force structure for the next 25 years, broken down into fighter, bomber, and attack; airlift and rescue systems; and other specialized systems. The charts indicate that the ACC will have to sustain primary combat systems for at least another 15 years until the F-22 and Joint Strike Fighter begin to enter the inventory in significant quantities.

---

[1] Most older aircraft, which were designed prior to the ASIP update in the 1970s that incorporated damage tolerance requirements, subsequently had a DADTA to define the inspections and modifications needed to protect structural safety.

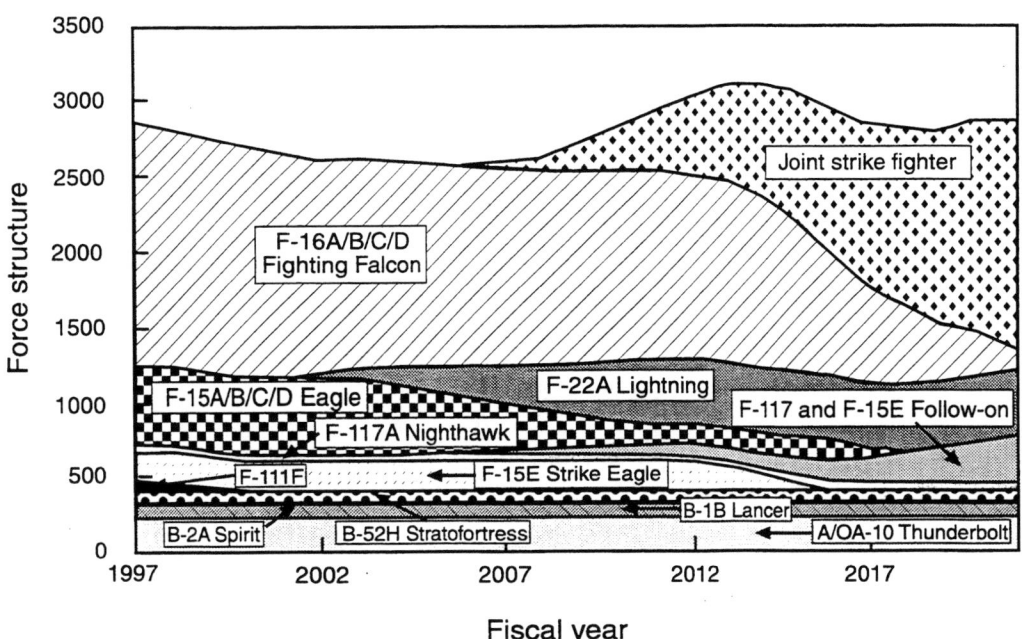

FIGURE 2-1 Force structure projection for the ACC fighter, bomber, and attack aircraft. Source: JACG (1996).

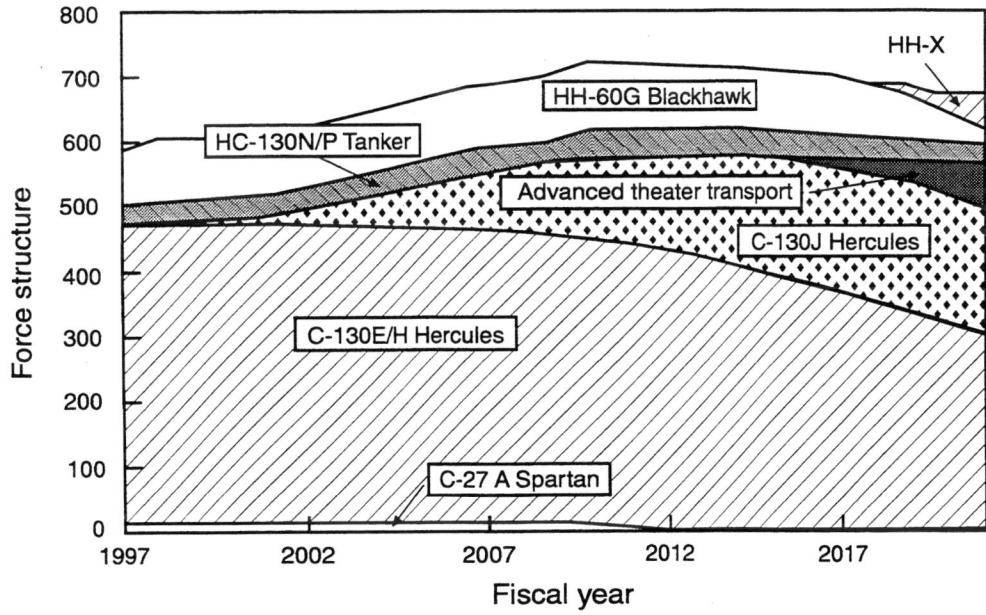

FIGURE 2-2 Force structure projection for the ACC airlift and rescue aircraft. Source: JACG (1996).

Air Mobility Command

The aircraft program chart for the AMC (Figure 2-4) reflects their core mission aircraft that support the air refueling and airlift missions. The C-141 will be retired in the near future and will be replaced by the C-17. The KC-135 and most of the C-5 forces, however, are being extended for most of the next 25 years.

Special Operations Command

As shown in Figure 2-5, the AFSOC employs a wide variety of different aircraft systems in small force sizes to support their mission of force application, mobility, and psychological operations. The AFSOC is currently modernizing their forces, replacing some aging helicopters with the CV-22 tiltrotor over the next ten years. The AFSOC has also recently introduced the

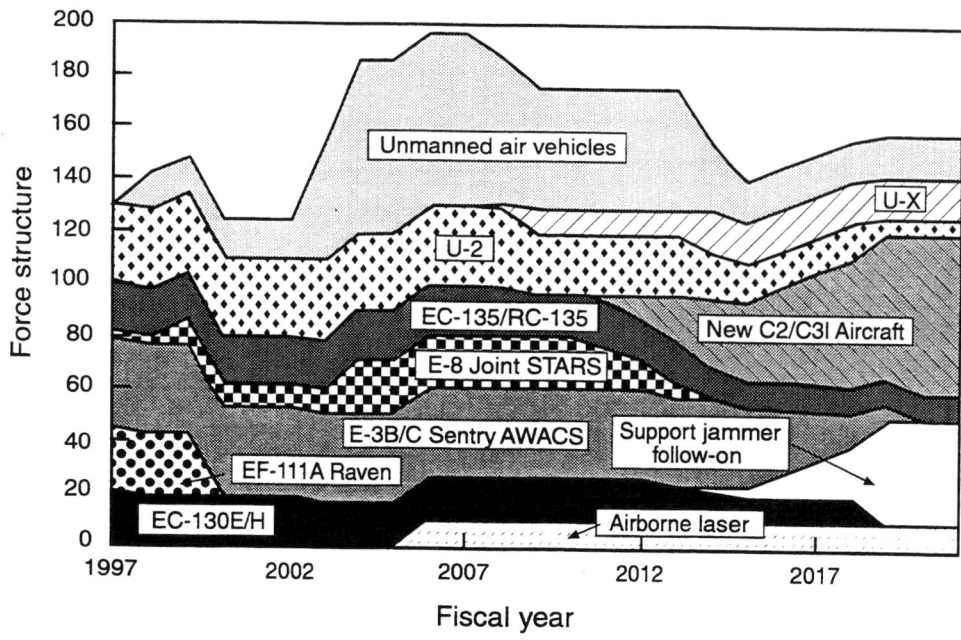

FIGURE 2-3 Force structure projection for other ACC aircraft. Source: JACG (1996).

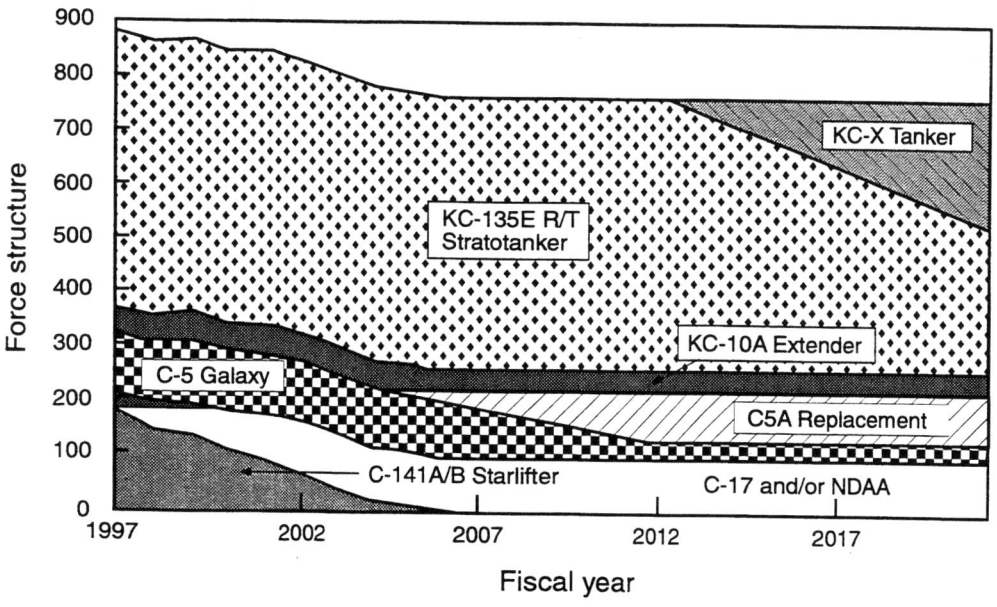

FIGURE 2-4 Force structure projection for AMC aircraft. Source: JACG (1996).

AC-130U gunship and the MC-130H. The AFSOC force presents some unique challenges in staying abreast of aging problems because of their specialized configurations and mission profiles.

Air Education and Training Command

The AETC operates a relatively new force for its pilot training activities, except for the T-38 and T-43, some of

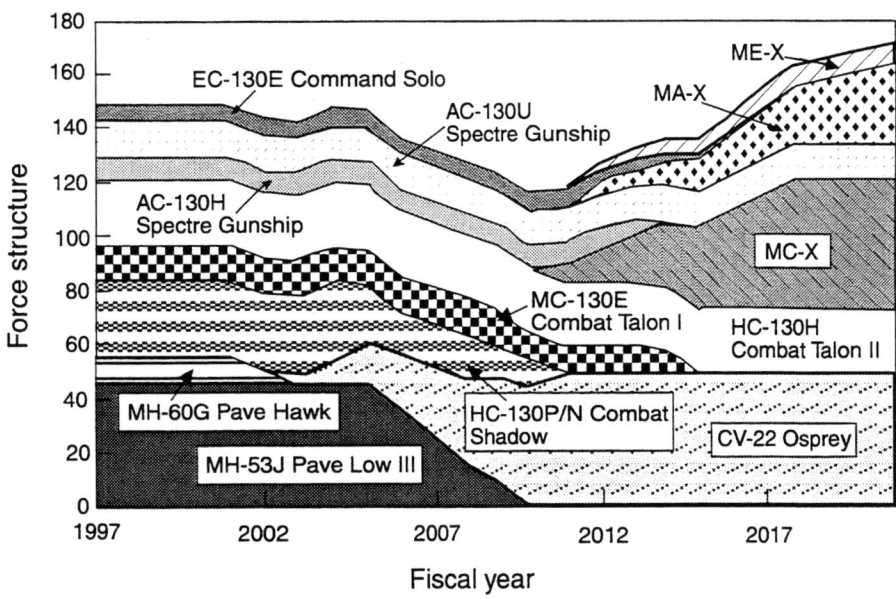

FIGURE 2-5 Force structure projection for AFSOC aircraft. Source: JACG (1996).

which will be over 50 years old by the year 2020. The aging T-37 is being replaced by the JPATS over the next ten years. The future force structure for AETC is shown in Figure 2-6.

ASSESSMENT OF THE FORCE STRUCTURAL MANAGEMENT PROCESS

Modernization Planning Process

The Air Force modernization planning process contains the essential elements for effective force structure planning and management at all levels. It links the Air Staff, major operating commands, system program directors, and the technology community into a comprehensive planning system. The effectiveness of the AFMPP depends on the timeliness and accuracy of information used to develop the plan. For this reason, it is critical that estimates of the economic service life of the aircraft weapon system be as accurate as possible and be progressively improved as new information becomes available. Continued efforts to employ and refine the AFMPP should ensure that structural economic life considerations are incorporated into the force structure planning.

Service Life Projection

Currently, there is no clear definition of all the elements that constitute the determination of structural economic life for aircraft systems, or a standard economic model to assist in determining when the costs of operating and maintaining the system reach a level that clearly warrant replacement. Lack of these tools frustrate the ability of Air Force planners to establish a realistic time table to phase out a current system and to begin planning for replacement systems. In addition, no comprehensive system exists for forecasting or assessing the total yearly operating and maintenance costs for an aircraft. An economic-based model to estimate the cost-effective service life would greatly facilitate force structure planning and give credibility to system replacement decisions and budget requirements. Given the long lead times (i.e., more than ten years) for replacement systems, along with their supporting technologies, timely and accurate service life forecasts for current aging systems are critical to maintaining force readiness. The committee believes that the development of an overall economic service life estimation methodology that integrates the time-dependent effects of structural deterioration with economic considerations is essential to force management.

ASIP and Postproduction Force Management

The primary focus of ASIP has been and continues to be to protect structural safety. ASIP was originated with the approval of General Curtis LeMay in 1958 as a result of a series of wing failures on B-47 bombers. However, it was not until the 1970s, with the introduction of damage tolerance requirements into ASIP and use of DADTAs of older aircraft, that the problem of unacceptably high aircraft losses due to structural fatigue failures was finally brought under control. Since this major revision to ASIP

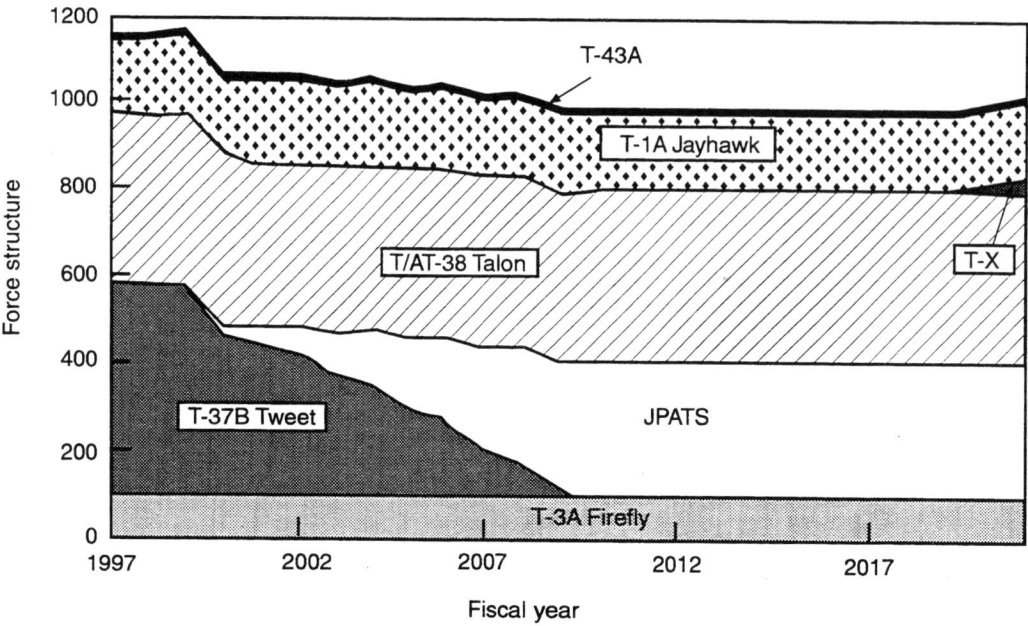

FIGURE 2-6 Force structure projection for AETC aircraft. Source: JACG (1996).

in the early 1970s the program has been a huge success. An indication of this success is that the failure rate for all weapon systems that are maintained using the damage tolerance approach is one aircraft lost due to structural reasons in more than ten million flight hours (Lincoln, 1997). This is two orders of magnitude less than the aircraft loss rate from all other causes.

It has been the implementation of the structural inspections and modifications that have been derived from the damage tolerance approach and applied to the aircraft weapon systems by the Air Force's logistics and operational support organizations that has so successfully protected the structural safety of Air Force aircraft for more than two decades. However, the committee is concerned that the extended use of old aircraft, coupled with the current trends of reducing military budgets and manpower, increased reliance on contracted maintenance, use of commercial design practices rather than military specifications, and possible complacency of Air Force management (because of the greatly reduced number of aircraft lost due to structural failures in recent years) may make this past success rather fragile. It is the committee's opinion that the effectiveness of the damage tolerance approach and its success in preventing structural failures has been dependent on a number of factors:

- rigid "cradle to grave" enforcement of ASIP by the system program offices at Wright-Patterson AFB and the system program directors at the air logistics centers (ALCs) as required by Air Force Regulation 80-13 and by the aircraft contractors as required by MIL-STD-1530A and the supporting military specifications, which have been required on aircraft weapon system contracts
- implementation of the IATPs to allow the maintenance program to account for the large variations and changes in usage that are commonplace in military combat aircraft, but virtually absent in normal commercial aircraft operations
- technical oversight of the DADTAs by an experienced standing Air Force committee up until the mid-1980s, plus periodic reviews of specific weapon systems by Air Force Scientific Advisory Board committees and by Division Advisory Group committees for the Aeronautical Systems Division
- development of competent ASIP managers and engineering support groups within the ALCs, that had the capability to perform damage tolerance analyses, monitor contractor's analyses, and keep the FSMPs and IATPs up to date
- sufficient funding of the DADTAs, IATPs, and the other ASIP support activities at the ALCs
- adequate R&D funding to address design, analysis, inspection, and maintenance and repair needs

Based on discussions with Air Force engineering and logistics personnel, the committee believes that the relatively recent acquisition reforms, budget and manpower reductions throughout the Air Force, and engineering-grade structure limitations at the ALCs have all adversely affected these factors. It will take aggressive actions by Air Force management and engineers to counter deterioration in capability and loss in the ASIP oversight that apparently has already occurred and to prevent further deterioration in the future. ASIP should continue to be enforced, and sufficient resources should be maintained to continue to track aircraft, keep the damage tolerance assessments up to date, and keep corrosion and stress corrosion cracking from becoming a structural safety issue. Also, sufficient R&D resources should be maintained to support and improve aging aircraft engineering, inspection, and maintenance and repair. Recommended R&D and engineering and management tasks are expanded on in Part II.

3

Current Structural Status of the Aging Force

When discussing the Air Force's aging aircraft, it is helpful to consider the Air Force-supported aircraft separately from the Air Force commercial-derivative aircraft, which typically use contractor logistics support (i.e., contractor logistics-support aircraft). Both are discussed in the following sections.

AIR FORCE-SUPPORTED AIRCRAFT

Table 3-1 summarizes data on aircraft age and planned future replacements for aging aircraft that are maintained by the Air Force.[1] During the course of this study the committee received briefings and written material on these aircraft. Some of the more significant structural problems encountered with these aircraft are discussed briefly below. Additional details on all the aircraft listed in Table 3-1 can be found in Appendix A. The B-2 bomber, the F-117 attack aircraft, and the C-17 airlifter are excluded because of their relatively recent introduction into the force. Also, because of time, budget, and technical considerations, the committee elected to exclude the H-1, H-53, and H-60 rotorcraft from this study. Helicopters are somewhat unique in that dynamic excitations in the rotor systems (i.e., causing combined high- and low-cycle fatigue) have been at the root of many past structural problems, and as such the committee suggests that this subject could best be addressed separately.

The Air Mobility Command's (AMC) airlifter and tanker aircraft listed in Table 3-1 were designed based on the fail-safe approach, and, as a result, the primary safety concern with regard to aging is the loss of this fail-safety from the onset of widespread fatigue damage (WFD).[2] In fact, both the KC-135 and the C-5A had their original lower wing surfaces replaced in the 1970s and early 1980s because of WFD, and the wings of the C-141 have more recently undergone extensive modification because of WFD (i.e., use of boron composite doublers to repair and prevent further cracking at the weep holes in the lower wing surface risers). In addition, some C-141s are now experiencing WFD in the lower wing surface spanwise slices. Risk analyses performed by the aircraft manufacturer have shown that these splices reach the onset of WFD at about 37,000 flight hours. As shown in Table 3-1, the current plan is to retire all of the C-141s within the next eight years; however, this plan was based on an aircraft retirement time of 45,000 flight hours. For any aircraft that must be flown more than 37,000 hours before retirement, extensive and burdensome inspections are required to protect the structural safety. These inspections involve inspecting more than 6,000 fastener holes per aircraft every 120 days. The committee is not aware of any data to indicate that the KC-135 or the C-5 will experience the onset of WFD in the near future. For example, a blue ribbon panel reviewed the KC-135 during 1996 and concluded that the current data indicate that the aircraft could likely be flown to beyond the year 2040 before encountering WFD. However, the panel recommended some additional actions to improve this estimate and emphasized the need to control the present corrosion and stress corrosion cracking problems.

The Air Combat Command's (ACC) fighter, bomber, and attack aircraft and the Air Education and Training Command's T-37 and T-38 trainer aircraft were either designed to be damage tolerant using the safe crack growth concept or were later analyzed on the basis of crack growth to establish safety limits and inspection requirements. This was accomplished during the durability and damage tolerance assessments (DADTAs) that were performed on these aircraft. Although some of these aircraft have some inherent failsafety resulting from redundancy in load paths and crack arresting features or because of battle damage requirements, they do not meet the fail-safe standards of the large transport aircraft. As such, with increasing age the primary threat to their structural safety is the growth in fatigue-critical areas and the potential of missing one or more of these areas. As noted in Chapter 4 and Appendix A, there already has been a significant increase in the number of critical areas in the F-16 since its introduction into service in 1979. Also, based on the number of cracking locations currently being reported in the T-38 and A-10 (see Appendix A), it appears that this is also true for these aircraft.

The T-38 is of particular concern because of its singleplank lower wing skin, its very small critical crack sizes (i.e., 0.20 to 0.40 in.), and the age of the aircraft in terms of both calendar years and flight hours. Wing failure and

---

[1] An exception is the U-2, which was developed for the government and maintained by the contractor.

[2] See Chapter 4 for a discussion of technical issues associated with aging aircraft, including widespread fatigue damage.

TABLE 3-1 Data on Force Status for Air Force-Supported Aircraft

Aircraft Operator	Aircraft Type	Current Age Data		Future Plans
		Years Since IOC[a]	Average Age (years)	
Air Mobility Command	Airlifter and Tanker Aircraft			
	KC-135	41	35	Retain 25+ years. No replacement identified
	C-5	28	18	Retire C-5A in 10–15 years. No replacement identified
	C-141B	32	29	Retire over next 8 years. Replace with C-17
Air Combat Command	Bomber and Attack Aircraft			
	A-10	20	15	Retain 25+ years. No replacement identified
	B-52H	36	34	Retain 25+ years. No replacement identified
	B-1B	11	9	Retain 25+ years. No replacement identified
	F-15	23	12	Retire in 5–20 years. Replace with F-22
	F-16	18	8	Retire in 10–25 years. Replace with Joint Strike Fighter
	Other Aircraft			
	C-130E/H[b]	36	20	Replace 1/3 over 5–25 years with C-130J
	E-3 (AWACS)	20	16	Retire in 17–25 years. No replacement identified
	E-8 (JSTARS)	N/A	15–20	Retire in 15–20 years. No replacement identified
	EC/AC-135	40	30–35	Retain 25+ years. No replacement identified
	U-2[c]	40	14	Retire in 15–25 years. No replacement identified
	EC-130E/H	36	20	Retire in 15–25 years. No replacement identified
	EF-111	30	29	Retire within next 4–5 years
Air Education and Training Command	Trainer Aircraft			
	T-37B	38	33	Retire in 2–12 years. Replace with JPATS
	T-38	36	29	Retain 25+ years. No replacement identified

[a]IOC: initial operational capability
[b]Operational control of the C-130E/H was recently transferred from the Air Combat Command to the Air Mobility Command.
[c]This aircraft was developed for the government and is maintained by the manufacturer rather than by an air logistics center.

aircraft losses occurred during the 1970s when these aircraft were put into severe roles. The T-38 was used in the lead-in-fighter (LIF) role[3] and the dissimilar air combat training role by the Tactical Air Command. The critical crack size that caused the wings to fail was also about 0.2 in. Since the 1970s it appears that the Air Force's San Antonio Air Logistic Center and their prime contractor, Northrop-Grumman, have done a good job of maintaining structural safety and preventing wing failures through the use of safety inspections,[4] structural modifications, design changes, and lower wing surface replacements. Additional full-scale wing fatigue testing has also been performed to identify critical areas in the new and modified structure. Crack growth analyses have performed to establish inspection requirements. Further design changes, wing replacements, and full-scale testing are anticipated by the San Antonio Air Logistics Center. The committee concurs that these changes undoubtedly will be needed if the aircraft are to remain in the inventory for 25 years or more.

Table 3-1 also shows several of the ACC's other aircraft that are used in various missions involving electronic combat, surveillance, intra-theater airlift, and tracking of enemy air and ground forces. Except for the U-2, the structures of these aging aircraft are predominantly of a fail-safe design, in which the threat to safety is the onset of WFD. The E-3 and the E-8 are both derivatives of the Boeing commercial 707 aircraft. However, the E-8 airframes are old commercial airframes that have been modified, whereas the E-3 were new airframes based on the 707 design. In fact, several of the E-8s have airframes that exceed the original design life goal (i.e., 20,000 flights) and, as indicated in Appendix A, a number of the aircraft are either at or are believed to be approaching the onset of WFD and will very likely require lower wing surface replacements in the near future.

The C-130 aircraft included in Table 3-1 (i.e., the C-130E/H and the EC-130E/H) have been in production for more than three decades. The E models were delivered

---

[3]Currently, the equivalent to LIF is called Introduction to Fighter Fundamentals or IFF.

[4]There was one additional aircraft lost (during the 1980s) due to a wing crack that should have been detected.

between 1961 and 1972 and they make up the majority of the active Air Force assets. The H model, which has been supplied since 1973, makes up the balance of the intra-theater airlift and EC-130 electronic combat capability. There have been a number of fatigue cracking and corrosion problems over the years that have led to the retirement of nearly all of the A models and outer wing replacements of the B and E models to the H configuration. There have also been numerous center wing replacements on the B, E, and H models. As pointed out in Appendix A, the major uncertainty about the C-130E/H airframe is the probable service life of the fuselage and the associated future structural maintenance needs. During 1996 an Air Force Structural Review Team looked at this issue and made several recommendations, including the teardown inspection of a high-time aircraft to look for evidence of WFD and the performance of a DADTA to determine future safety inspection requirements.

CONTRACTOR LOGISTICS-SUPPORTED AIRCRAFT

The Air Force's contractor logistics-supported (CLS) commercial-derivative aircraft are listed in Table 3-2. These aircraft range in average age from about 3 years to more than 30 years. In some cases, such as the KC-10 and the C-27, there were DADTAs performed under Air Force guidance. For the E-4 and the C-18, the Air Force had the manufacturer modify their damage-tolerance-derived inspection intervals for anticipated Air Force use. For the most part, the aircraft listed in Table 3-2 have been designed and certified to Federal Air Regulation requirements (e.g., FAR Part 25 for the large transport aircraft and FAR Part 23 for utility and commuter aircraft) and are contractor maintained to commercial standards. One exception is the C-27, which is a later model of a military transport aircraft originally developed in the 1970s by Aeritalia for the Italian Air Force (i.e., the G222TCM).

Because of time and budget limitations, the committee did not attempt to review each of these aircraft with regard to corrosion, fatigue, and stress corrosion cracking histories or their specific Air Force use spectra. However, it has been pointed out to the committee that many of these aircraft have very low utilization rates compared with their commercial counterparts and in many cases are being flown to operational spectra comparable with those flown in commercial operation. Table 3-3 shows a utilization comparison between the large Air Force aircraft shown in Table 3-2 with their commercial counterparts. With the exception of the C-9, the data in this table support the position that, for those commercial

TABLE 3-2 Air Force Commercial-Derivative Aircraft Using Contractor Logistics Support[a]

Air Force Designation	Commercial Designation	Quantity	Average Age (years)	Operator(s)[b]
E-4	Boeing 747-200	4	23	ACC
VC-25	Boeing 747-200	2	7	AMC
T-43	Boeing 737-200	13	24	ACC and ANG
C-137	Boeing 707-100/300	6	21	AMC
C-18	Boeing 707-323	6	N/A	AFMC, ACC, USAFA
C-22	Boeing 727-100	3	32	ANG
KC-10	McDonnell Douglas DC-10-30F	59	13	AMC
C-9	McDonnell Douglas DC-9-30	23	26	AMC, USAFE, PACAF
C-12	Beechcraft Super King Air 200	37	17	AFMC, PACAF, AETC
T-1A	Beechjet 400A	156	3	AETC
C-21	Learjet 35A	76	13	All commands
C-23	Shorts 330	3	13	AFMC
C-26	Fairchild SA227 Metroliner	40	5	ANG
C-27	Alenia G-222 Model 710A[c]	10	5	ACC
C-20	Gulfstream II, III, IV	13	10	AMC and USAFE
UV-18	Dehaviland DHC-6 Twin Otter	2	20	USAFA
E-9	Dehaviland DHC-8	2	N/A	ACC
T-3	Slingby T67M260 Firefly	112	3	AETC and USAFA

[a] Excludes six types of glider aircraft, two small Cessna aircraft used by the Air Force Academy, and newer aircraft not yet in the inventory (e.g., C-32).

[b] Operators: ACC (Air Combat Command), AMC (Air Mobility Command), AETC (Air Education and Training Command), ANG (Air National Guard), AFMC (Air Force Matériel Command), USAFA (United States Air Force Academy), USAFE (United States Air Forces in Europe), PACAF (Pacific Air Forces).

[c] Not a commercial aircraft, but a military transport originally built for the Italian Air Force.

TABLE 3-3 Comparison between Utilization of Air Force CLS Aircraft and Commercial Equivalents

	Air Force CLS Aircraft[a]		Commercial Aircraft[a]						
				Number of Flights			Flight Hours		
Aircraft	Flights[b]	Hours	Aircraft	Average	High	Design Goal	Average	High	Design Goal
E-4	7,500–11,000	8,000–10,000	747	~10,000	~32,000	20,000	~40,000	~95,000	60,000
VC-25	N/A	~2,500	747	~10,000	~32,000	20,000	~40,000	~95,000	60,000
T-43	10,000–15,000	16,000–18,000	737	~20,000	~85,000	75,000	~22,500	~80,000	60,000
C-22	51,000–55,000	57,000–59,000	727	~35,000	~72,000	60,000	~47,000	~78,000	60,000
C-18	13,000–44,000	33,000–62,000	707[c]	~20,000	~37,000	20,000	~40,000	~90,000	60,000
VC-137	8,000–24,000	7,000–52,000	707[c]	~20,000	~37,000	20,000	~40,000	~90,000	60,000
KC-10	1,400–2,500	6,300–13,000	DC-10	N/A	~36,000	42,000	N/A	~90,000	60,000
C-9	11,600–51,200	11,000–44,600	DC-9	N/A	~99,000	40,000[d]	N/A	~79,000	30,000[d]

[a] Approximate data as of 1995 for the commercial aircraft and 1996 for the Air Force CLS aircraft.
[b] Except for the KC-10 and C-9, the data for the Air Force CLS aircraft reflect number of landings, which may be slightly larger than number of flights.
[c] There are 57 707 aircraft remaining in commercial use in the world. There are none registered in the United States.
[d] Contractor revised values to 102,000 flights and 78,000 hours based on retest and tear-down inspection of high-time commercial aircraft.

derivatives that the Air Force purchased new (i.e., the E-4, VC-25, T-43, and KC-10), the use in terms of both total flight hours and number of flights is low compared with their commercial counterparts and thus would not be expected to experience the onset of widespread fatigue cracking in many more years of operations. The high-time C-9s on the other hand have exceeded the contractor's original design life goals in terms of numbers of flights and total flight hours, but is still considerably less than the 102,000 flights and 78,000 hours that the aircraft manufacturer has now verified by retest and tear-down inspection of a high-time commercial aircraft. Whether or not these revised numbers apply directly to Air Force use is still a question that the committee believes should be investigated.

In addition, some of the large CLS aircraft were not purchased new and had a significant amount of commercial use prior to being modified for Air Force use. As seen from Table 3-3 there are C-22 (727), C-18 (707), and VC-137 (707) aircraft, in which the total number of flights or flight hours are, in some cases, close to or exceed the commercial design life goals. As indicated in Appendix A, the 1996 partial tear-down inspection of the lower wing skin and stringers of a commercial 707-300C aircraft with 57,382 flight hours and 22,533 flights revealed over 1,500 fatigue cracks. Most of these cracks were small, but some were large enough to indicate a high risk of a potential structural failure. This would indicate that the high-time C-18s and VC-137s may be approaching the onset of WFD in their wings and should be investigated further. Although the Air Force has indicated that some of the VC-137s will be replaced in the near future by new C-32 (i.e., 757) aircraft, the committee is not aware of any plans to replace the C-18. The C-22 (727) wings are less susceptible to WFD than the C-18 or VC-135 because of lower stress levels, better stringer materials, and an improved fastener system, but the C-22 fuselage is a potential fatigue cracking concern that also needs to be investigated further.[5]

With regard to the smaller utility and commuter class aircraft and the C-27 military transport listed in Table 3-3, the committee lacks the information necessary to make any judgments about their structural health and probable longevity. It is noted that several of the aircraft types, which comprise more than 300 aircraft, have very low average ages (i.e., the T-1A, T-3, C-26, and C-27) and thus one would not anticipate fatigue cracking very soon. On the other hand, a fundamental shortcoming in the current FAR Part 23 requirements is that there is no requirement for the aircraft to be designed to be damage tolerant (i.e., either fail-safe or safe crack growth). This shortcoming has been recognized by the Federal Aviation Administration and some commuter aircraft manufacturers, and changes to both the design rule and the supporting advisory circular are currently in process. Nevertheless, these types of aircraft that are currently in service have a large variance in their damage tolerance capabilities. In many cases they have single load-path structures, and the failure of a

---

[5] Early model 727 aircraft, which used cold-bonded fuselage lap splices, can experience the onset of WFD very early (i.e., in less than 30,000 flights) if the bonding becomes ineffective and preventive modifications have not been made.

single member could result in the loss of the aircraft. To minimize the possibility of such an occurrence, the Federal Aviation Administration has had a team of structural experts conducting damage tolerance surveys of selected aircraft in the commuter fleet over this past year. It is the team's intent to identify aircraft with the highest potential for structural failure, define further damage tolerance analysis requirements, and consider other actions that may be necessary to minimize the potential for failure. It would appear prudent for the Air Force to initiate a similar effort for the CLS utility- and commuter-sized aircraft listed in Table 3-2. These independent surveys or reviews would be to assess the current structural health of each type of aircraft, determine the need for a more detailed DADTA, improve corrosion control, and determine if an economic service life assessment of the aircraft is warranted.

4

Technical Issues and Operational Needs

In this chapter the primary aircraft aging mechanisms are discussed briefly in relation to their impact on aircraft structural health and longevity, and the associated technical issues and operational needs are identified. This is followed by brief discussions of technical issues and needs in the areas of nondestructive evaluation and maintenance and repairs. The focus of this chapter is on the degradation mechanisms for aluminum alloy airframe structures, which are predominant in current aging aircraft. Issues and recommendations concerning future aging aircraft, including composite primary structure are included in Chapter 10. The issues and needs identified in this chapter are the basis for the recommended engineering and management actions and near-term and long-term research and development presented in Part II of this report.

The three primary mechanisms that can affect the structural health and longevity of the Air Force's metallic aircraft structures are

- corrosion
- stress corrosion cracking
- fatigue cracking (including low-cycle and high-cycle fatigue)

CORROSION

Corrosion of airframe structure is the single, most costly maintenance problem for Air Force aging aircraft (SAB, 1994). Corrosion can occur in a variety of nonexclusive forms, including uniform or general corrosion, galvanic corrosion, pitting corrosion, fretting corrosion, crevice (filiform and faying surface) corrosion, intergranular (including exfoliation) corrosion, and stress corrosion cracking (ASM, 1987). Because of the potential structural effects, stress corrosion cracking is considered in greater detail in the following section.

Corrosion of aging aircraft results from a combination of factors, including

- the use of aluminum alloys and tempers that are more susceptible to corrosion than currently available alternatives
- inadequacy or deterioration of corrosion protection systems
- exposure to various corrosive environments (e.g., humid air, saltwater, sump tank water, latrine leakage)

Control of corrosion is predicated on effective prevention, detection, and repair methods. Despite the best intentions of prevention and control practices, the complete elimination of corrosion is virtually impossible. In aging aircraft structures, corrosion protection and control systems deteriorate over time. The major concern with the deterioration of corrosion protection systems for aging aircraft structure is the resulting increase in maintenance costs because corrosion damage that is identified must be repaired. Based on current experience, this practice of identifying and repairing corrosion damage has been adequate for maintaining the integrity of aging structures. However, because corrosion damage is typically found by visual inspection techniques, and a fair amount of corrosion damage occurring on older Air Force aircraft is hidden from direct view, a significant amount of corrosion can remain undetected. Also, there can be a wide variation in extent and severity of corrosion damage among similar aircraft because of differences in environmental exposures and in the amount and type of maintenance that the aircraft may have received.

The different types of corrosion damage exhibit different characteristics and potential consequences with respect to both detectability and structural consequence. For example, exfoliation corrosion (severe intergranular corrosion where the buildup of corrosion products causes flaking and surface blisters) and pitting corrosion can be detected readily, depending on the accessibility of the damaged surface. Although these corrosion forms are evident as surface deterioration, they may not be found if the surface is inaccessible to visual inspection. Intergranular corrosion that propagates along grain boundaries away from exposed surfaces may be indistinguishable from the surface, challenging the reliability of nondestructive inspection (NDI) methods (Mindlin et al., 1996).

Undetected corrosion can progress significantly before being observed, leading to (1) increased maintenance costs and time in the depot for maintenance or (2) an increased risk that corrosion, in the presence of other forms of damage, may cause a more significant decrease in damage tolerance than otherwise estimated. This is discussed later in this chapter. Although corrosion can be very costly to repair, corrosion by itself has not yet caused any structural failures that have resulted in the loss of an Air Force aircraft. This is because corrosion has been detected and repaired before it could become a flight safety problem.

Corrosion prevention begins—or should begin—during engineering design with proper selection of materials and manufacturing processes. As part of structural maintenance programs, the commercial aircraft industry has developed provisions to upgrade corrosion resistance through the use of substitute materials and heat treatments (e.g., more corrosion-resistant 7050, 7150, or 7055 alloy for 7075, stress corrosion- and exfoliation-resistant T-7X tempers for 7XXX-series aluminum alloys), improved protective finishes and corrosion-preventive compounds (CPCs),[1] and incorporation of design features such as drainage and sealing to prevent corrosion. However, similar engineering guidelines that provide advice on materials and processes having better corrosion resistance than the original materials and processes have not been formally developed for Air Force aircraft.

A panel chartered under the ad hoc committee of the Air Force Scientific Advisory Board concluded that a reduction of the relative humidity to 30–40 percent would significantly reduce the corrosion of stored aircraft and that existing dehumidification and storage systems appeared to be adequate (SAB, 1996). The report describes equipment and logistics relevant to dehumidified storage and discusses successful dehumidification programs in the other U.S. services (e.g., Marine Corps A-6E, Navy SH-60B, and Army CH-47D) and internationally (e.g., Swedish and Danish air forces). Since the Scientific Advisory Board study, work has been accomplished to validate the practicality of dehumidified short-term and long-term storage facilities for a range of military hardware using desiccant wheel technology (Cannava, 1997). It is not yet clear that dehumidification will be cost-effective for Air Force aircraft.

The committee believes that, if costly component repair and replacement are to be avoided, much more emphasis should be given to early detection of corrosion and implementation of effective corrosion control and mitigation practices. A practicable and more cost-efficient strategy for dealing with corrosion damage of airframe structures is needed to effectively guide prevention, control, and force management decisions for aging aircraft. The most important operational needs include

- environmentally compatible protective coatings to replace the hazardous materials being phased out (e.g., chromates)

- generalized use of CPCs and development of CPCs that can be applied on external surfaces and that will penetrate and protect unsealed joints and around fastener heads on older aircraft structures[2]
- guidance for the application of advances in alloys and processes offering improved corrosion protection
- improved techniques to discover and roughly quantify hidden corrosion without requiring disassembly of the aircraft
- classification of corrosion severity, similar to current commercial aircraft practice, to provide guidance to maintenance actions
- improved understanding of the probable rates of corrosion and corrosion trends for specific operational aircraft for use in planning maintenance actions
- dehumidified storage of aircraft or dehumidification of susceptible areas of particular aircraft

With improved detection methods and the implementation of improved corrosion prevention and control actions, the committee does not believe that physical corrosion damage per se will limit the structural life of Air Force aircraft.

STRESS CORROSION CRACKING

Stress corrosion cracking (SCC) is an environmentally induced, sustained-stress cracking mechanism. Early metallic aircraft built from thin aluminum sheet experienced few stress corrosion problems. The occurrences that did occur (e.g., pressed-in bushings) were generally diagnosed quickly and solved with little fanfare. However, in the post-World War II era, increasing numbers of stress corrosion problems appeared as a result of the introduction of high-strength 7XXX-series aluminum alloys and the growing use of integrally stiffened structure. The latter entailed installation of shaped components machined from thicker starting stock, which in turn introduced heat treat and assembly residual stresses unanticipated in the original design process. Designers and the aluminum industry gradually learned how to reverse the alarming rise in SCC incidents through an awareness of causes, through the identification of susceptible alloys, and through the introduction of tougher materials with improved corrosion resistance, improved surface preparation processes, and reduced residual stresses.

The results from comprehensive industry surveys on SCC service failures conducted during the late 1960s and early

---

[1] Internally-applied CPCs have been used as a maintenance material for commercial aircraft since the late 1960s. These materials, generally high molecular weight petroleum products that displace moisture from the potential corrosion cell, have been effective in both delaying the progression of incipient corrosion and in preventing corrosion from forming. CPCs are a critical part of maintenance programs to prevent and control corrosion and are finding increased use in new aircraft, especially in lower fuselage areas.

[2] The report of the Materials Degradation Panel of the Air Force Scientific Advisory Board (SAB, 1996) suggests that MIL-C-81309, a water-displacing, thin, soft film CPC can be used effectively on the aircraft exterior if reapplied periodically.

1970s formed the backbone of current Air Force practices for management of SCC (ASM, 1987; Spiedel, 1975). In general, it was found that the majority of aluminum airframe structures documented as failing by SCC were manufactured from the high-strength 7XXX- and 2XXX-series aluminum alloys that contain Al, Cu, Zn, and Mg. Of these alloys 7075, 7079, and 7178 in the peak strength T6 condition and alloy 2024 in the naturally aged T3 condition contributed to more than 90 percent of the reported aluminum SCC failures. A number of SCC problems have also been observed in high-strength steels. The SCC associated with service failures was observed to be characteristically intergranular, making early visual detection somewhat difficult.

SCC is generally exacerbated by residual tensile stresses remaining from material heat treatment or fit-up, but can also be triggered by operational loads and forces from the buildup of corrosion by-products. Aircraft designers are well aware of the reduced mechanical properties of forgings and thick plate materials in the short-transverse grain direction compared with those in the longitudinal grain directions. As a result, structural components are usually designed so that the primary load paths are parallel to the principal grain direction. In this case, the elongated grain boundaries are parallel to, rather than normal to, the applied operational stresses. Fortunately, when SCC occurs parallel to applied operational stresses, cracks often can be very large (e.g., as much as several inches long) before they become a flight safety problem.

When components containing SCC are discovered, they are either replaced with new components fabricated from more-resistant materials and tempers or they are repaired. Both replacement and repair are often quite difficult and costly (e.g., repairing or replacing sections of large fuselage bulkheads). The need for replacement and repair of components due to SCC could be reduced, or at least delayed, with appropriate maintenance actions. Important operational needs include

- improved environmental protection systems (e.g., improved CPCs and surface finishes) to reduce the corrosion rate of susceptible components
- modified manufacturing practices that reduce exposed end-grain and residual stress effects that exacerbate SCC in large structural components
- improved repair procedures, ranging from structural repairs to the restoration of corrosion-retarding finishes
- reevaluation of SCC-susceptible structural components to identify potential safety problems; such evaluations could be done during an Aircraft Structural Integrity Program (ASIP) durability and damage tolerance assessment of the aircraft

With continued vigilance, improvement in prevention and control procedures, and replacement of susceptible components with corrosion-resisting alloys with minimized residual stresses, the committee believes that SCC problems in older Air Force aircraft can be managed and that SCC need not be a life-limiting damage mechanism.

FATIGUE CRACKING

Unlike corrosion and SCC, which the committee believes can be controlled and thus would not, in themselves, physically limit structural life, fatigue cracking is a direct result of aircraft use (i.e., load or stress cycles) and will eventually occur in all aircraft. Both low-cycle fatigue (typically due to flight maneuver and gust loading) and high-cycle fatigue (due to vibratory excitation from aerodynamic, mechanical, or acoustic sources) are of concern. Also, the potential accelerating effects of corrosion damage on fatigue cracking must be recognized and accounted for when estimating life limits and determining safety inspection intervals.

Low-Cycle Fatigue

Safety Limit and Economic Life Limit

Fatigue cracking and failures resulting from the growth of cracks from preexisting flaws or defects (introduced during material processing or manufacturing) have occurred quite early in the operational life of an aircraft. In single load-path structures, such as those used in many older fighter aircraft, the structural failures have often led to loss of the aircraft. Because of these failures, the Air Force extensively revised ASIP in the early 1970s to include damage tolerance design and test requirements. These requirements were incorporated in MIL-STD-1530A (DOD, 1988) and MIL-A-83444 (DOD, 1987) and required that the structure be designed with the a priori assumption that it contains the maximum probable-sized initial material or manufacturing flaw or defect located in the most critical areas of the structure. The time required for a crack to grow from this initial flaw size to the critical size (i.e., size at failure) was defined as the safety limit for the structure. The aircraft was not allowed to fly beyond the safety limit without a careful inspection (and repair or modification if necessary). In addition, repeat inspections were required at one-half the time required for the flaw to grow from the size that is barely undetectable by NDI methods to critical size. If the structure was not inspectable or was difficult to inspect, the aircraft operator had only three options once the safety limit was reached: modification, structural replacement, or retirement. For new aircraft, a design goal is to avoid in-service structural safety inspections by selecting stress levels and materials such that the calculated safety limits exceed the expected design life of the aircraft when operated according to the design service use spectrum. The primary problem with this design goal has been that actual service use has very often been different, and sometimes much more severe, than the original design spectrum.

Deviation from original design spectra have occurred as a result of increased weight and increased frequency of high-load occurrences per flying hour. To a large part, the increased weights have been caused by additional armaments and electronic systems, and the increase in high-load occurrences are the result of changes in tactics. Both have been caused by the change in military threats.

Experience indicates that it is important that, for aircraft having single load-path structures, (1) all fatigue-critical areas and likely root causes for fatigue crack initiation be identified, (2) critical loading conditions be defined, (3) safety limits be established based on actual use conditions, and (4) in-service safety inspections be performed. It is also important that the Air Force maintain the individual aircraft tracking program so that inspection intervals can be adjusted based on actual use and understanding of failure modes.

As aircraft age increases, it is more likely that fatigue cracks will be detected in critical areas and that additional areas of the structure will become critical and will require more inspections, repairs, and modifications. Eventually, the safety inspection, repair, and modification costs, when combined with other maintenance costs (e.g., for corrosion, stress corrosion, and wear), and the reduced availability of aircraft will become so burdensome that it will be more cost effective to replace the aircraft than to continue to maintain it. The Air Force calls this the economic life limit for the aircraft.

As discussed in Chapter 2, the committee believes that the prediction of when a given aircraft will reach its economic life limit is a crucial issue in future force planning. Although various attempts have been made to estimate economic life limits for various aircraft (e.g., the F-4E, KC-135), the Air Force has not yet developed a method for estimating such limits that accounts for all of the relevant cost factors.

From the standpoint of maintaining safety of aircraft that contain single load-path structures, the most important issue is ensuring that all critical areas of the airframe and flight load conditions have been identified. Currently, the Air Force relies on full-scale fatigue test results, past service experience, and various analyses (e.g., stress, fracture mechanics, failure modes, and environmental effects) for the identification of critical areas.

Corrosion Effects on Safety Limits

The Air Force recognizes that crack growth rates may be influenced by environmentally induced corrosion that in turn would affect the safety limits and repeat inspection intervals. In an attempt to account for this effect, data on the rate of crack growth have been developed for materials in environments that are believed to be appropriate (e.g., in high humidity, salt spray). Crack growth is limited to the time to reach the K_{ISCC} (i.e., the threshold stress intensity for SCC) in this environment to avoid superposition of fatigue and stress corrosion crack growth rates.[3]

Although the damage tolerance requirements and the approach for accounting for potential environmental effects have served the Air Force well over the past two decades, the committee continues to be concerned that, as structures age, as corrosion protection systems continue to deteriorate, and as materials corrode, there may be effects that have not been adequately considered. Specific corrosion concerns or issues that could affect safety limits and repeat inspection intervals include

- the influence of corrosion on applied stresses and stress intensity factors resulting from material thinning and local bulging of thin sheet due to buildup of corrosion products
- the potential influence of corrosion on material mechanical properties (toughness, strength, elongation) resulting, for example, from the absorption of hydrogen by the metal during the corrosion process
- the potential influence of corrosion and corrosive environments on crack growth rates below the threshold for SCC

Widespread Fatigue Damage

Although failure that is due to fatigue crack growth from an initial material flaw is of lesser concern in the larger transport and tanker aircraft, which generally have been designed to be fail-safe either through the use of multiple load paths or through the use of crack arrest features,[4] there is serious concern about the loss of fail-safety due to the onset of widespread fatigue damage (WFD) initiating from normal quality structural details. This is of particular concern for older aircraft or aircraft that have been flown under a more severe use spectrum than that for which it was designed.

The onset of WFD in a structure is characterized by the simultaneous presence of small cracks in multiple structural details; where the cracks are of sufficient size and density, the structure can no longer sustain the required residual strength load level in the event of a primary load-path failure or a large partial damage incident. Multiple-site damage and multiple-

---

[3]In practice, the K_{ISCC} cutoff is often ignored, with negligible effect on safety limits, when K_{ISCC} is high with respect to the critical stress intensity. This is the case for most aluminum alloys, except in the short-transverse grain direction of thick plate, extrusions, or forgings.

[4]Failure of the structural member or load path or large partial failure will not result in loss of aircraft because of the second line of defense provided by the alternate load paths or structure surrounding the large partial failure.

element damage are subsets of WFD, where there are multiple cracks either in the same structural element or in adjacent structural elements. When the onset of WFD occurs, the airframe (or major component of the airframe) has reached the operational life limit. To preclude unsafe operations once the onset of WFD occurs, flight restrictions or groundings are the only options until the affected structure is modified, replaced, or the aircraft is retired.

To assist in future force structure planning, it is necessary to be able to predict when the onset of WFD will occur (i.e., estimate when a sufficient number and sizes of cracks will degrade the residual strength of the structure to below the fail-safe design level) and to assess if WFD concerns will affect a significant part of the force structure. To predict the onset of WFD, it is necessary to (1) predict accurately the residual strength of the structure after encountering the primary damage (e.g., from a discrete source) with various sizes of small WFD cracks in the adjacent intact structure and (2) predict when these cracks will occur. Although there has been considerable effort over the past few years to develop analytical models that enable prediction of the residual strength of fuselage longitudinal lap splices (as a result of the 1988 Aloha Airlines 737 accident), there has been much less effort directed toward structural configurations typical to military aircraft (e.g., thick chordwise wing joints, plate and stringer wing configurations, shiplap spanwise wing splices, large pressure door hinges, loading ramp hooks, and others). The development of analytical residual strength models for the various structural configurations and materials that are typical of military aircraft is an important technical issue to address in the characterization of WFD.

The Air Force has been almost entirely dependent on using the results from detailed tear-down inspections of full-scale fatigue test aircraft, or actual high-time fleet aircraft, to predict when the small WFD cracks will occur. Tear-down inspections are required because current fatigue analyses are inadequate to predict accurately the initiation and growth of very small flaws, although there has been some promising research in this area. The primary shortcomings of using only full-scale fatigue test results to predict crack initiation are (1) full-scale fatigue test results are not usually representative of the actual operational load spectrum, and (2) the potential influence that environmental exposure may have on the crack initiation process is neglected. Although the tear-down inspection of actual fleet aircraft is the most reliable basis for determining the onset of WFD, it entails the destruction of one or more aircraft (or major portions of aircraft) and comes too late to provide data for force planning.

The difficulty in analytically predicting the initiation and growth of small cracks arises, in part, from the potential that several different mechanisms (e.g., corrosion, fretting, microstructural defects, residual stress) could influence crack initiation at any given structural location. The identification of which crack initiation mechanism is most likely to lead to a fatigue crack is problematic; the likelihood is remote that all of the controlling parameters that contribute to crack initiation will be modeled rigorously in the formulation of a mechanics boundary value problem. Analytical methods based on small crack theory have been shown to be quite accurate in predicting total fatigue life of laboratory test specimens by using fracture mechanics and initial crack sizes determined from the characterization of microstructural defects. However, this approach represents only one possible initiation mechanism. Based on the status of current research, the most promising analytical approach to predict the behavior of other initiating mechanisms is to use an equivalent initial flaw (EIF) size determined from coupon and structural element tests. In fact, small crack theory uses the EIF approach, but determines the EIF from microstructural features that are characterized with microscopy rather than calculating the EIF size from fatigue test data. The committee believes that a comprehensive EIF-based fracture mechanics approach, including simulative experimental methods for characterization of feature demographics and the prediction of initiation and growth of small fatigue cracks, is vital to the development of an analytical capability to allow the prediction of the onset of WFD. Such a capability requires the development of an EIF database, correlated with full-scale structural test articles, for cracks that initiate because of fretting, very small defects, scratches, dings, and corrosion damage.

Among the greatest NDI challenges is to develop methods that can reliably, rapidly, and cost effectively determine, without fastener removal or disassembly, if an aircraft has widespread fatigue cracking. Inspection for WFD is difficult because the crack sizes that can significantly degrade the structure are, in most cases, very small, and there is a very large number of structural details (e.g., fastener holes) that need to be inspected. The specific crack sizes that must be detected depend on the specific structural configuration, the materials used, and the design stress levels. However, it is not unusual for cracks as small as a few hundredths of an inch (e.g., 0.04 to 0.10 in.) to be sufficient to degrade the structure's fail-safe residual strength to below the design operating load level. Although there are existing nondestructive methods (e.g., eddy current and ultrasonic methods) that can find such small cracks, they are very tedious, time-consuming, and costly when applied to large areas of an aircraft. Also, the methods are much more limited, less sensitive, and less reliable when the cracking occurs in the inner layer of a wing or fuselage joint or in an interior structural member such as stringers or spar caps.

High-Cycle Fatigue

High-cycle fatigue occurs when a structure is exposed to high-frequency load cycles from aerodynamic, mechanical, and acoustic sources. The amplitude of these load cycles are not as high as those experienced during normal flight maneuver loading, but the frequency is high and thus the structure can be subjected to a very large number of damaging load cycles in a very short period of time. As a result, low-amplitude cycles can be sufficient to cause rapid fatigue crack initiation in unflawed structure or cause the stress intensity associated with even very small initial flaws in the structure to exceed the threshold for fatigue crack propagation (i.e., the K_{th}). In such cases, cracks will propagate very rapidly to critical size and failure will ensue.

High-cycle fatigue failures have occurred during service operations in a number of military aircraft in the recent past (e.g., acoustic fatigue of B-1 horizontal tail, buffet load damage of F-15 and F/A-18 vertical tails), even though the airframes were subjected to extensive structural testing over the full range of expected service conditions. Generally an attempt is made during the design to identify all possible sources of excitation. These sources are either eliminated or avoided, or the structural response is reduced substantially through design modifications. However, problems can arise later in service for a number of reasons:

- changes in the response of the structure (e.g., natural frequency) that are due to changes resulting from wear, corrosion, loose fasteners, repairs, and low-cycle fatigue crack growth
- changes in aircraft use, which in turn causes changes in the loading environment (i.e., new or magnified aerodynamic or acoustic excitation sources)
- changes in aircraft configuration, which in turn generate new sources of aerodynamic excitation (i.e., new weapon stores, pods, antenna, or moldline changes that modify air flow and shock impingement locations)

When a reduction in stiffness changes the natural frequency of the structure or a particular component, the primary need is to understand the cause of the stiffness loss and determine the tolerable damage state. This is generally done by a combination of analysis and testing of the component and by a comparison with loads data from the particular airframe. If the margins between the natural frequencies and the driving force are not large enough, analysis must be performed of potential stiffness increases to ensure that the problem is not simply moved to another location.

When changes in aircraft use cause high-cycle fatigue damage, an examination of aircraft load history, along with ground vibration and flight test data, must be used to determine the dynamic load history of the component. In some cases, load monitoring or sensing may be necessary to determine the relationship of local loads to the maneuver loads that are recorded by the aircraft's flight data recorder. In many cases, the dynamic loads occur under extreme environmental conditions (e.g., temperature extremes, fluid exposures), necessitating remotely interrogated sensors for in-flight loads measurement.

Detailed analyses, and perhaps flight data measurements, are required to assess changes in excitation loads and stiffness when configuration changes are the primary cause of high-cycle fatigue damage. Extreme care must be taken to ensure that repairs or modifications intended to reduce dynamic effects do not cause further harm. The repairs themselves can cause changes in air flow and local stiffness that result in dynamic loading problems at other locations.

The committee believes that dynamic loading and the resulting high-cycle fatigue is a key aging aircraft issue as well as an initial design issue, particularly for high-performance combat aircraft. The key technical issues include

- identification, reduction, or elimination of sources of dynamic excitation
- passive and active methods to reduce the response of aircraft structures
- measurement and characterization of the threshold for fatigue propagation (K_{th}) values for airframe materials, including the applicability of long crack thresholds to small crack behavior
- in-flight monitoring of changes in dynamic loading

NONDESTRUCTIVE EVALUATION

The development of nondestructive evaluation (NDE) technology for aging airframe structures is driven by structural requirements and cost considerations. Proper application of currently available NDE technology can offer significant improvements in diagnostic capabilities and provide characterization of the damage necessary to develop effective structural repairs. In addition, NDE methods protect structural safety by detecting, providing quantifiable characterization, and screening fatigue cracking, stress corrosion cracking, and corrosion conditions that are, or could become, a flight-safety concern. However, field practices and implementation of NDE methodology to meet many aging aircraft problems are inadequate and often inconsistent with current technical capabilities (e.g., field systems often do not take advantage of technology used during production of the structural components).

The committee has identified critical inspection needs based on the important aging mechanisms. The most important needs include

- detection of fatigue cracks under fasteners; the inability to detect cracks beneath fasteners can result in unmanageably short inspection intervals for fatigue-critical structures with small critical crack lengths
- detection of small cracks associated with WFD for cracks as small as a few hundredths of an inch

- techniques to discover and quantify hidden corrosion without disassembly of the aircraft
- detection and characterization of cracks and corrosion in multilayer structures
- detection of SCC in thick sections

Reliability is one of the most important characteristics of an effective NDE method (Berens, 1989; Cowie, 1989; Panhuise, 1989; Rummel, 1989). NDE inspection is a statistical process that depends on the inherent variability of many features including flaw size, orientation, distance of flaw from surface, surface roughness, and variations in material properties. A frequently used measure of the reliability of a NDE system is the probability of detection (POD) which is a conditional probability defined as the probability that a flaw with given characteristics will be found in an inspection. Obviously, a requirement for an efficient NDE inspection is a high value of POD for the particular flaw and geometry involved in the inspection.

Often, NDE methods must be able to quantitatively identify defects (e.g., in terms of size and location) that can affect structural safety. The maximum allowable defect size at a given location within the structure is determined by the structural analyst based on analyses or tests that demonstrate that such a defect will not grow to critical size within a specified period of operational use. It must then be shown that the selected NDE method will detect, within a specified POD, defects larger than these allowable sizes with a minimum number of "false calls." If this is not achievable, a solution must be developed so that the detection requirements are compatible with detection capabilities (e.g., allowable flaws or defect sizes could be increased by significantly increasing the required frequency of inspection).

The effective maintenance of aging aircraft is vitally dependent on implementing effective NDE methods. In addition to the specific needs identified above, there is an overarching need to improve the cost and time effectiveness of NDE inspections. One of the key barriers to implementing force-wide improvements to NDE methods is the huge scale of the task (Hagemaier and Hoggard, 1993). Although the specific needs are focused primarily on two fairly generic problems (i.e., detection and characterization of cracks and corrosion), each application involves different component geometry and structural configuration, requiring revalidation and qualification of NDE methods. Current practice requires the establishment of NDE reliability (POD), usually with an empirical test matrix of various inspections and specimens with known flaw types, sizes, and locations. To provide a statistically significant sample that accounts for the range of conditions likely to be encountered in a force-wide application, the size of the sample set would be enormous and costs would be prohibitive. The committee believes that new approaches and tools, including techniques to predict the response and reliability of new NDE methods, are needed to address this problem.

STRUCTURAL MAINTENANCE AND REPAIRS

An effective airframe structural maintenance program evaluates (1) sources (root causes) of structural deterioration; (2) susceptibility of the structure to each source of deterioration; (3) the consequences of deterioration to continued airworthiness; (4) the effectiveness of detection methods in finding structural deterioration, taking into account inspection thresholds and intervals (NRC, 1996a); (5) the effectiveness of the repair in restoring load-carrying capability and the effect on the integrity of surrounding structure; and (6) the effectiveness of prevention and control measures to mitigate existing and anticipated problems. When structural deterioration is detected in the maintenance program, a decision must be made to either repair or replace the affected components. The primary damage mechanisms to be considered for aging aircraft (discussed in preceding sections of this chapter) include corrosion, SCC, low-cycle fatigue (including WFD), and high-cycle fatigue.

In the case of corrosion, the primary issue involved in the restoration of corroded structure (when thickness loss that is due to corrosion does not necessitate structural repair) is the removal and reapplication of protective finishes to prevent further corrosion. The chief technical issues include

- environmentally compliant finish removal techniques to replace grit blast (grit disposal problems) and chemical methods (volatile organic releases and toxic substances)
- environmentally compliant surface preparation processes and finish materials that reduce or eliminate releases of heavy metals (e.g., chromium, cadmium) and volatile organic compounds
- evaluation of finish system durability and life

Repair of damage resulting from in-service degradation mechanisms such as fatigue, SCC, corrosion (when thickness loss requires structural repair), and discrete source damage (e.g., foreign object impact, handling damage, lightning attachment) is a critical maintenance activity. Generally, repair of aged structure consists of reinforcement doublers that are bolted or bonded over the damaged area. Bolted repairs are generally preferred for commercial aircraft because they are relatively simple to perform and minimize the time that the aircraft is out of service. However, bolted repairs introduce stress concentrations at fastener holes and tend to add considerable weight. Recent Air Force efforts have emphasized bonded composite patch repairs, even though the repairs are more complex and time consuming to design and install. Bonded repairs avoid stress concentrations from drilled holes, are more readily conformable to complex shapes, and provide more efficient load transfer at lower weight compared with bolted repairs. The primary technical issues for structural repairs include

- analysis methods and design practices for repairs
- material and process selection and optimization, including surface preparation, lamination, and bonding processes, and specification infrastructure for bonded repairs
- life prediction and inspection intervals for the repaired structure
- maintaining the damage tolerance of the repaired structure

Another important issue in the maintenance of aging systems is the replacement of components that are fabricated using alloys and processes that are susceptible to deterioration, especially corrosion and SCC. Improvements in commercial aluminum alloys, tempers, process controls, and fabrication methods could be beneficial if guidelines were provided to the logistic centers for materials and process substitution decisions (including performance and cost tradeoffs). Commercial airframe manufacturers have been active in updating obsolete materials and process specifications in their core design and manufacturing practices and in their maintenance programs (Goranson, 1997).

II

RECOMMENDED STRATEGY AND OPPORTUNITIES FOR NEAR-TERM AND LONG-TERM RESEARCH

One of the primary objectives of this study was to identify an overall strategy that addresses the Air Force's aging aircraft needs. From the discussion of the aging aircraft problem in Chapter 2, including the assessments of the force management process, it is apparent to the committee that the recommended overall strategy *must* encompass several engineering and management issues as well as the near- and long-term research opportunities. The committee believes that there are a number of engineering tasks that do not require additional research and that should be accomplished in the near future.

Also, to be effective, the strategy must address the three Air Force objectives that are noted in Chapter 1:

- identify and correct structural deterioration that could affect safety of flight
- prevent or minimize structural deterioration that could become an excessive economic burden or adversely affect force readiness
- predict, for the purpose of future force planning, when the maintenance burden will become so burdensome, or the aircraft availability so poor, that it will no longer be viable to retain the aircraft in the inventory

To provide a comprehensive approach that addresses these challenges, the committee recommends that the Air Force adopt a three-pronged strategy that includes (1) near-term engineering and management tasks, (2) a near-term R&D program, and (3) a long-term R&D program. This overall strategy is illustrated in Figure II-1.

Engineering and management tasks are near-term actions (within three to five years) to improve the maintenance and force management of aging aircraft. Each of the three aging aircraft challenges are shown on the left side of the figure connected to the primary engineering and management task that addresses each challenge. It should be noted, however, that this is not exclusively true. For example, the engineering task of obtaining improved corrosion control programs is connected to the challenge to minimize maintenance costs and improve readiness, since corrosion is currently the major contributor to maintenance costs and does not normally affect structural safety. However, corrosion could become a safety issue if not brought under control. Likewise the primary focus of the engineering tasks of updating durability and damage tolerance assessments, force structural maintenance plans, and tracking programs is to protect the structural safety, but they also impact maintenance costs and force readiness. The task of estimating the economic service life of an aircraft weapon system involves both engineering and management; engineering predictions of structural deterioration need to be coupled with a number of cost and operational considerations to arrive at the most probable time that the Air Force should plan on replacing the system. The last three tasks in Figure II-1 deal primarily with the Aircraft Structural Integrity Program and postproduction force management concerns discussed in Chapter 2 and further expanded in Chapter 5.

With the exception of the technology transition task, which is considered to be a continuous effort throughout the life of a weapon system, all of the near-term engineering and management tasks are shown to extend over a five-year period. Also, it is envisioned that some of the tasks should have periodic updates about every five years as indicated in the figure. The background justification and specific recommended actions for each of the eight engineering and management tasks are included in Chapter 5.

Supporting the near-term engineering and management tasks are the near-term R&D efforts that the committee believes should be performed under the direction of the Air Force's laboratories either in-house or by supporting contractors and academic institutions. Also, the Air Force laboratories should utilize the results from complementary near-term R&D efforts that are under the direction of other government agencies (i.e., the National Aeronautics and Space Administration, the Federal Aviation Administration, and the Navy). Figure II-2 illustrates the basic elements of both the near-term and the long-term R&D programs. The near-term program includes those efforts that reasonably can be expected to provide results that will assist in the performance of the near-term engineering tasks during the next five years. The long-term R&D program includes those efforts that the committee believes will take longer than three to five years to achieve a mature technology that could be adopted by industry or the Air Force aircraft maintenance organizations, but nevertheless should be initiated now, or continued if they already have been initiated. These efforts are typically higher

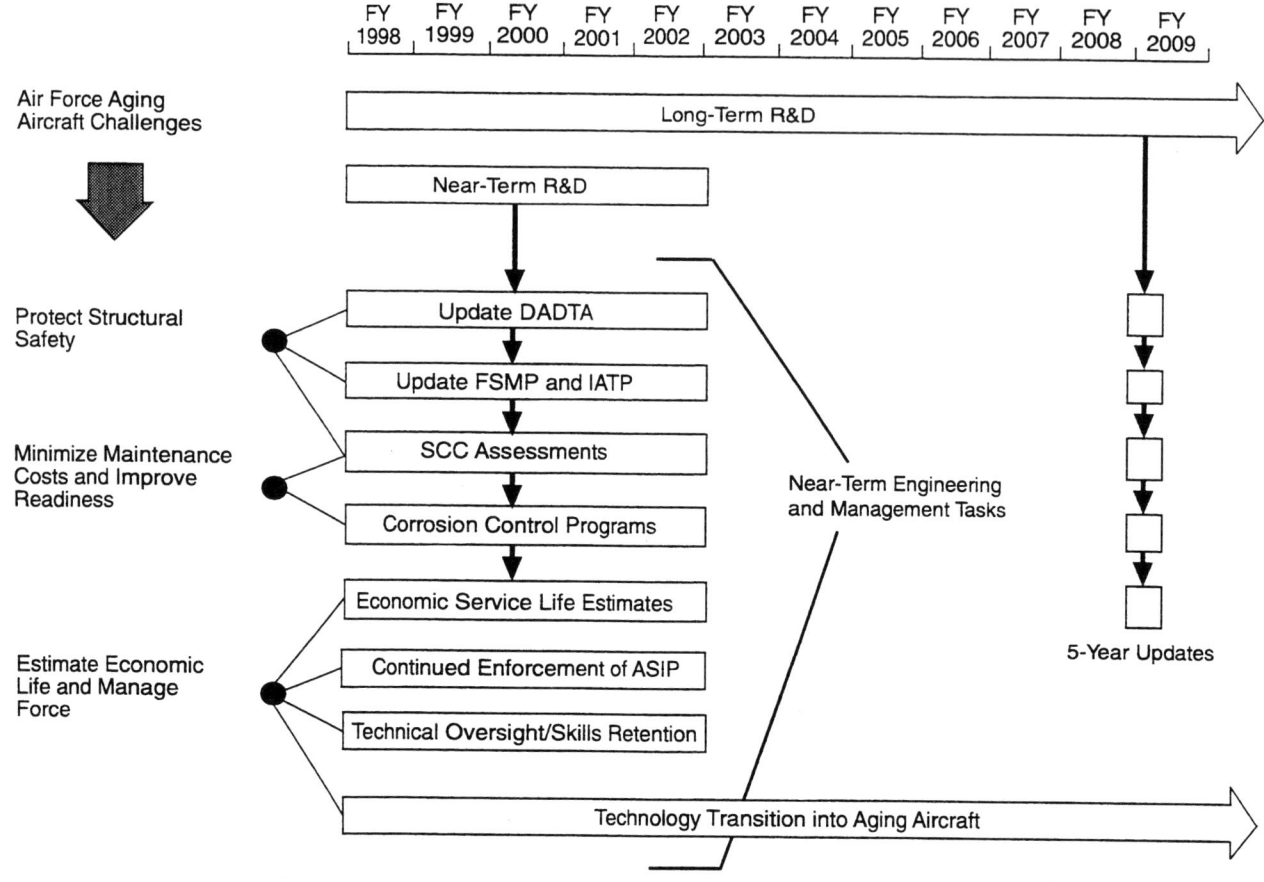

FIGURE II-1 Recommended overall strategy to address Air Force aging aircraft challenges. Strategy includes near-term engineering and management tasks and near-term and long-term R&D programs.

risk than the near-term R&D efforts, but the potentially high payoff justifies their pursuit.

Included in Part II are descriptions of recommended near-term engineering and management tasks; assessments of current and planned research administered by the aging aircraft research program (detailed assessments are contained in the committee's interim report [NRC, 1997]); identification of near-term and long-term research opportunities in the areas of fatigue (low-cycle fatigue, high-cycle fatigue, and environmental effects), corrosion and stress corrosion cracking, and inspection and maintenance technology (nondestructive evaluation and maintenance and repair); and prioritization of recommended research.

Although the investigation of structural aging phenomena is an inherently interdisciplinary endeavor, for convenience the recommended research is presented separately for individual topical areas. Chapters 6 (fatigue), 7 (corrosion and stress corrosion cracking), and 8 (nondestructive evaluation and maintenance) describe R&D opportunities focused on the aluminum structures that dominate the current aging aircraft problems. Chapter 9 provides prioritization of the near-term and long-term research recommendations. Finally, Chapter 10 describes issues related to composite primary structures that are becoming more common on newer aircraft that represent the next generation of aging aircraft.

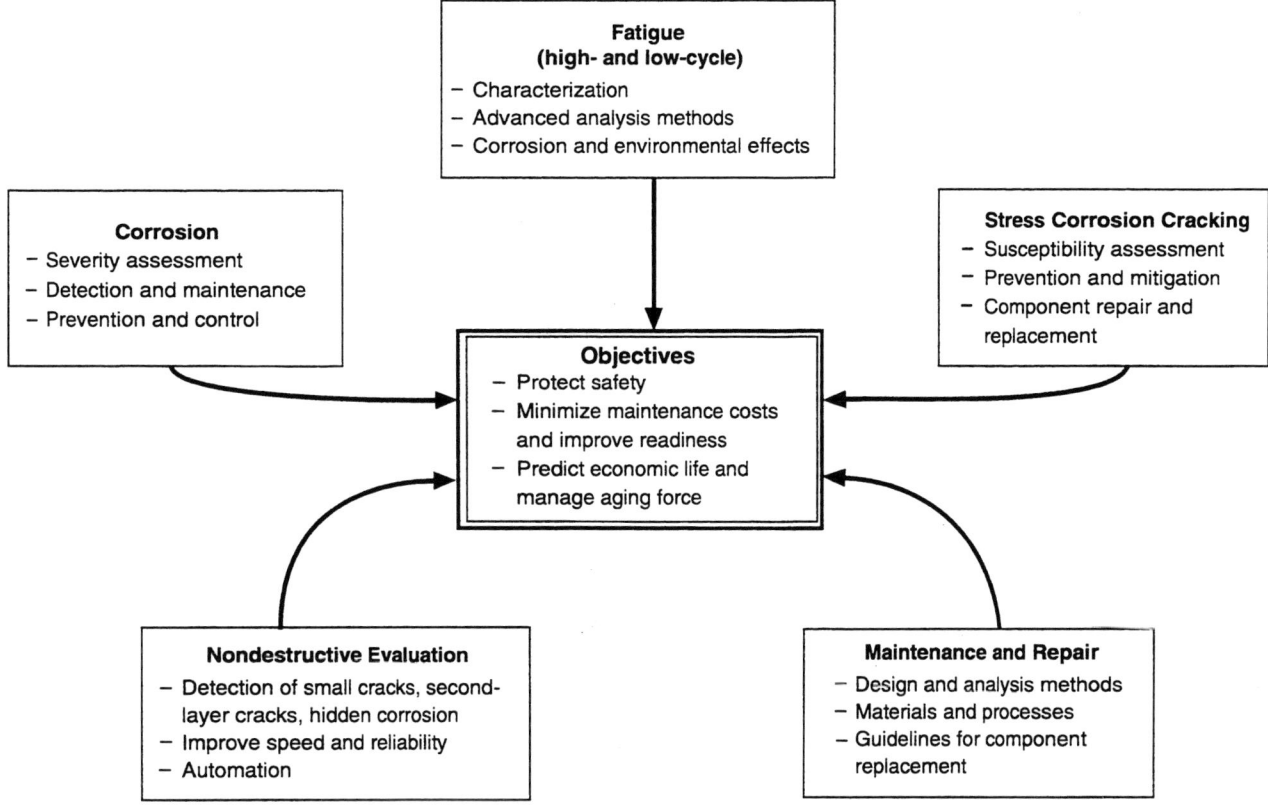

FIGURE II-2 Basic elements of the recommended near-term and long-term R&D programs.

5

Engineering and Management Tasks

UPDATE OF DURABILITY AND DAMAGE TOLERANCE ASSESSMENTS

As noted in Chapter 4, a number of aircraft failures resulting from fatigue crack growth from preexisting flaws or defects, which were introduced during material processing or manufacturing, caused the Air Force to extensively revise their Aircraft Structural Integrity Program (ASIP) in the early 1970s to include damage tolerance requirements. These requirements were defined in MIL-A-83444 (DOD, 1987) and MIL-STD-1530A (DOD, 1988) and were incorporated into the designs of the new aircraft then under way (e.g., the B-1A, F-16, and A-10). However, to protect the structural safety and assess the durability of the vast majority of Air Force aircraft that were not designed to these requirements, the Air Force and the aircraft contractors performed durability and damage tolerance assessments (DADTAs) on the aircraft models that were already in the operational inventory. By the early 1980s DADTAs had been performed on the F-4C/D/E, A-7D, C-5A, C-141, F-111, B-52D, E-3A, F-5E, T-38, T-37/A-37, KC-135, SR-71, T-39, KC-10, C-130, and F-15. Also, because of changes in use conditions, the durability and damage tolerance of both the A-10 and F-16 had to be revisited after only a short time in operational service.

From the standpoint of safety, the most important outputs from these assessments were the identification of fatigue-critical areas, the determination of safety limits for these areas, and the development of safety inspection requirements. In addition, for some of the larger transport aircraft, estimates were made of the onset of widespread fatigue damage (WFD) and risk analyses were performed (e.g., on the C-5A, KC-135, and C-141). Where appropriate, lower-bound estimates were made of the major component modification or replacement times and modification options were defined.

The overall approach or methodology used in conducting the DADTAs is illustrated in Figure 5-1. As can be seen in this figure, the four primary tasks in the assessments are (1) the identification of fracture-critical areas;[1] (2) the development of the operational stress spectra for these areas; (3) an assessment of initial flaw distributions and/or the maximum probable initial flaw sizes; and (4) the determination of the safety limits, inspection intervals, and, for fail-safe designs, the estimated onset of WFD. The results were then used to update the individual aircraft tracking programs and the force structural maintenance plans for the aircraft, both of which are key elements of ASIP.

Air Force-Supported Aircraft

To obtain improved visibility of the actions that will be necessary to protect the structural safety of the Air Force's aging aircraft listed in Table 3-1 throughout their projected operational lives and to obtain the best estimates as to when the aircraft will likely be facing the economic impacts of major modifications or replacements, the committee strongly recommends that the DADTAs of these aircraft be updated periodically. In general, an update about every five years is appropriate.

The urgency to perform these updates varies among the different aging aircraft types, depending on several factors: (1) whether the aircraft structure is designed to be fail-safe or is largely of a single load-path design, where missing a critical area could lead to the loss of an aircraft; (2) whether a replacement aircraft has been identified and the older aircraft are being phased out of the inventory; (3) the extent and nature of fatigue cracking problems the aircraft are currently encountering; and (4) whether there has been a recent independent review of the aircraft and corrective actions are already under way. Table 5-1 summarizes these different factors for each of the Air Force's aging aircraft types shown previously in Table 3-1. Also shown in Table 5-1 is the committee's assessment of the priority that should be assigned to performing the DADTA update for each type of aircraft. Those of greatest concern, based on the highest potential for structural safety problems, were given a number 1 priority and those with the least immediate concern were given a number 3 priority. However, it is recommended that the DADTA update be performed on all of the aircraft within the next five years and updated at approximately five-year intervals.

The committee recognizes that the level of effort involved in performing these updates will vary significantly between the different types of aircraft as a function of aircraft complexity, variations in use, the numbers and types of cracking

---

[1] If rapid crack propagation and part failure could lead to the loss of the aircraft, it is defined as a fracture-critical area.

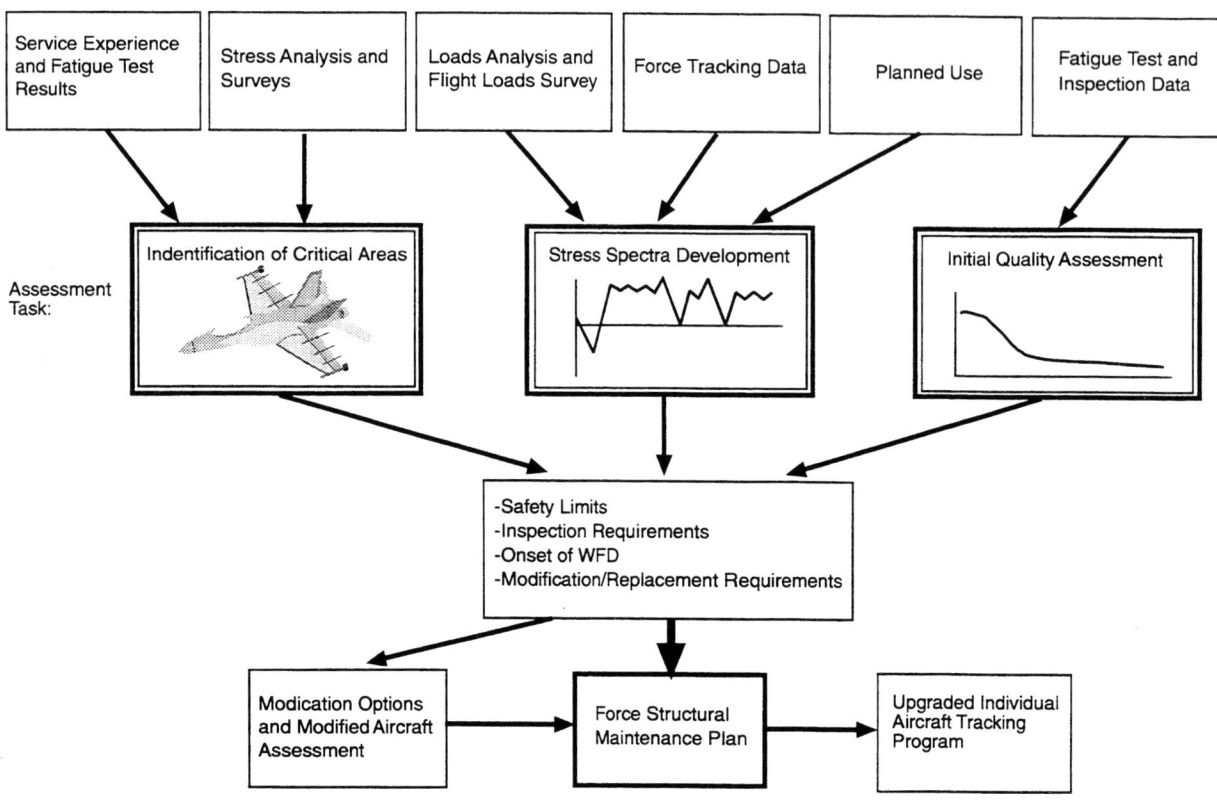

FIGURE 5-1 Overall approach to durability and damage tolerance assessments.

TABLE 5-1 Prioritization of DADTA Update Needs for Air Force-Supported Aircraft

Aircraft	Fail-Safe Design	Additional Years in Inventory	Replacement Aircraft Identified	Current Fatigue Cracking	Recent Structural Review[a]	Review Actions Under Way	Priority
KC-135	yes	25+	no	limited	yes	yes	3
C-5A	yes	10–25	no	no report	no	no	2
C-141B	yes	0–8	yes (C-17)	yes	yes	yes	3
A-10	no	25+	no	yes	no	no	1
B-52H	no	25+	no	yes	no	no	3[b]
B-1B	no	25+	no	yes	yes (horizontal tail)	yes	2
F-15	no	5–25	yes (F-22)	limited	no	no	2
F-16	no	10–25	yes (JSF)	yes	yes (fuselage bulkhead)	yes (bulkhead)	1
C-130E/H	yes	25+	some (C-130J)	limited	yes (fuselage)	unknown	2
E-3 (AWACS)	yes	17–25	no	limited	no	no	3
E-8 (JSTARS)	yes	15–20	no	yes	yes (wings)	unknown	2
EC-135	yes	25+	no	limited	yes	yes	3
U-2[c]	no	25+	no	unknown	no	no	1
EF-111	no	<5	no	limited	no	no	none[d]
T-37B	no	0–12	yes (JPATS)	limited	no	no	3[e]
T-38	no	25+	no	yes	no	no	1

[a]Within the past three years.
[b]The lower priority is because a DADTA update was performed in 1995.
[c]This aircraft was developed for the government and is maintained by the manufacturer rather than by an air logistics center.
[d]Based on the assumption that all aircraft will be retired in less than five years as planned.
[e]DADTA is currently being performed by Southwest Research Institute. Update suggested within five years.

problems encountered, and how well the different air logistics centers (ALCs) and the airframe contractors have been performing the applicable ASIP tasks on a continuing basis during the aircraft's past operational use. As a minimum, the effort may merely require a summary of available data (e.g., critical areas, safety limits, inspection requirements, estimates of the onset of WFD, estimated future modification and replacement times, and possible future fatigue test needs) and a detailed review by the proposed Aging Aircraft Technical Steering Group discussed later in this chapter. For other aircraft it will require further identification of critical areas, stress spectra development, crack growth calculations and tests, and perhaps some tear-down inspections and/or full-scale fatigue testing.

Contractor Logistics-Supported Commercial-Derivative Aircraft

In a similar manner to the criteria for Air Force-supported aircraft (previous section), priorities are suggested for contractor logistics-supported commercial-derivative aircraft. In addition to the criteria described in the previous section for Air Force-supported aircraft, the experience with the commercial-equivalent aircraft can be taken into account.

The KC-10 and C-27 have previously had DADTAs. It is recommended that they be updated within the next five years. Because there is no immediate safety concern, a priority 3 is suggested.

The E-4, T-43, and C-9 have average ages of 23, 24, and 26 years with plans to keep them in the inventory for many more years. It is recommended that the Air Force form an independent team to review these aircraft. This team should consist of a small number of structures and materials experts chartered to assess the current condition of the aircraft, review the current use spectra, and determine if the current contractor database is sufficient to estimate the onset of WFD and the probable major component modification or replacement times or if DADTAs should be performed. The committee suggests that these reviews be performed within the next five years. Because of the much higher use, it is recommended that the C-9 be addressed first. A priority 2 is suggested for the C-9 and priority 3 for the E-4 and T-43.

The C-18, C-22, and the VC-137 aircraft have quite high utilization times. There is some concern about the possible onset of WFD for the C-18, C-22, and possibly the VC-137. Thus, the committee recommends that an independent structures review be conducted by a team of structures and materials experts in the near future. If the high-use VC-137s are replaced by the C-32, they of course could be dropped from the review. Because of the potential safety implications, a priority 1 is suggested for these reviews.

For the utility and commuter class commercial-derivative aircraft (i.e., the C-12, T-1A, C-21, C-23, C-26, C-20, E-9, UV-18, and T-3), the committee recommends that the Air Force initiate damage tolerance surveys, by a small team of structures and materials experts, similar to those conducted during this past year by the Federal Aviation Administration (FAA) on a number of other types of aircraft in this size class. These surveys should provide a preliminary assessment of the aircraft's damage tolerance, current structural health and estimated longevity, and the potential need for a detailed DADTA. The surveys should be conducted first on the older aircraft or aircraft where structural problems may have already been identified.[2]

UPDATE OF FORCE STRUCTURAL MAINTENANCE PLANS AND INDIVIDUAL AIRCRAFT TRACKING PROGRAMS

The fourth and fifth tasks of the Air Force's ASIP (shown in Table 2-1) deal with force management. It is here that the results of design, analysis, and full-scale test activities in the previous parts of ASIP (including subsequent DADTAs) come together to define the specific actions that must be taken to protect the safety of the individual aircraft and allow for the timely and cost-effective structural modifications. The two key force management activities in ASIP are the development of the force structural maintenance plan (FSMP) and the individual tracking program (IATP).

Force Structural Maintenance Plan

During the initial design, the intent was to minimize the amount of structural maintenance that would be needed throughout the life of an aircraft, assuming that the aircraft is used as planned (i.e., it is flown to the design use spectrum). However, full-scale fatigue testing to the design spectrum will uncover critical areas missed during design and analysis, which then necessitates additional damage tolerance analysis, in-service safety inspections, and perhaps in-service modifications. It is the definition of when, where, how, and the estimated costs of these inspections and modifications that constitute the basis for the initial FSMP.

Recognizing that the actual service use of military aircraft often differs from the original design use spectra, ASIP requires that a loads/environment spectra survey be conducted during the first two or three years of operational service to obtain actual use data that can be used to update the original design spectrum. These surveys generally consist of instrumenting 10 to 20 percent of the fleet and using a

---

[2]For example, a potential fatigue cracking problem has been reported (by the Air Force's contractor logistics support office) to exist in some C-12 wing spars. This is a potential safety concern, since the structure is not a fail-safe design.

multichannel recorder (or more recently, microprocessor systems) to record such data as vertical and lateral load factor; roll, pitch, and yaw rates; roll, pitch, and yaw accelerations; altitude; mach number; rudder and aileron position; and selected strain measurements. These data are then used to generate a new baseline operational spectrum, and new damage tolerance analyses are performed to update the safety inspection and modification requirements with the results added to the FSMP. This updated FSMP then forms the basis for planning and scheduling the structural fatigue maintenance for the overall aircraft weapon system. The damage tolerance analysis should be updated and the results used to update the FSMP any time that there are significant changes in use; when operation is extended beyond the original life goal; or new analysis, test, or service experience indicate a growth in the number of fatigue-critical areas.

Individual Aircraft Tracking Program

In addition to the force-wide baseline operational use spectra being different from the original design spectra for military aircraft, the individual aircraft use within the force may be either more or less severe than that represented by the baseline spectrum. These variations from the baseline spectrum can be quite large, particularly for the high-performance combat type aircraft. Accordingly, the Air Force has included the requirement for individual aircraft tracking as part of the ASIP.

The IATPs for the various types of aircraft within the Air Force inventory vary with regard to data acquisition and processing procedures. For the larger tanker, transport and bomber aircraft (e.g., the KC-135, B-52, and C-141), where the excursions in the flight spectra are relatively small, flight logs and pilot use forms (i.e., Air Force technical order form 16 and tactical maneuver supplemental forms) have been found to be satisfactory to track the aircraft use. For the fighter and attack aircraft the use of counting accelerometers and VGH (velocity; ground range and height) recorders were commonplace in the past, but are limited because they are not able to accommodate critical areas of the structure that are sensitive to asymmetrical loading. The use of multichannel recorders (e.g., the MXU-553), which record many more flight parameters, overcomes this limitation. More recently, the older tape systems are being replaced (as funding will allow) by microprocessor systems, further expanding data capture. Computerized methods have been developed and are used to reduce the measured flight data and to adjust the crack-growth-based damage rates and inspection intervals for each of the critical areas in the airframe for individual aircraft use. As the aircraft ages, the number of critical areas and inspections increase. When this happens, the IATPs must be updated to accommodate these changes.

Although there has been some discussion about upgrading the Air Force's IATP to track potential corrosion damage and/or corrosive environments as well as fatigue damage, the committee believes that the application of sensor devices and data analysis and processing equipment in existing aircraft is currently impractical because of the large number of aircraft involved, the large sizes of affected areas in the aircraft most prone to corrosion damage (i.e., the large transport, tanker, and bomber aircraft), and the cost and intrusiveness of system installation. However, developments in multifunctional chemical and physical sensors (NRC, 1995), microelectromechanical systems, and smart diagnostics do provide some hope that long-term research in on-board health monitoring can be productive.

Following the completion of the updates of the DADTAs, which were recommended above, the committee recommends that

- the inspection and modification requirements in the FSMPs be updated to reflect any changes in the baseline operational spectra and any additional critical areas that were identified, which in turn will increase the inspection requirements and possibly necessitate new modifications
- the IATP for each aircraft weapon system be updated to reflect additional critical areas that need to be tracked plus any changes in sensors, recording equipment, or analysis procedures that may be deemed necessary to protect the structural safety of the aircraft. In particular, the Air Force should push for the force-wide use of the microprocessor-based recorders because of their improved reliability and the expanded data capture.

STRESS CORROSION CRACKING ASSESSMENTS

Although the environmental protection measures and material substitutions to eliminate corrosion-susceptible materials that take place as part of an aircraft's corrosion prevention and control program (CPCP) also apply to the prevention of stress corrosion cracking (SCC), there are some unique aspects about SCC that make this structural deterioration mechanism much more dangerous than other forms of corrosion. Thus, the committee believes that SCC deserves special attention. Stress corrosion cracks are characteristically intergranular and can occur with little or no evidence of corrosion products and as a result are often difficult to detect visually. Although they generally have not caused flight-safety problems, because of their orientation with respect to the applied flight stresses (see Chapter 4), this cannot always be considered to be a certainty. If large in-plane stress corrosion cracks or delaminations go undetected they could cause a loss in shear strength and trigger failure modes other than the tensile mode normally associated with crack propagation. Also, in thick sections (e.g., complex machined fittings) where there may be irregular grain flows and three-dimensionally applied stresses, it is often difficult to predict if a stress corrosion crack will turn normal to the largest component of stress and result in a tensile fracture.

For fail-safe structural designs, a part or component failure caused by a stress corrosion crack is much less of a concern than in safe crack growth designs because of the second line of defense provided by the surrounding intact structure. In fact, over the years there have been many part failures caused by SCC in both commercial and military aircraft. When this occurs, the parts are generally replaced, ideally with new parts made from more-stress-corrosion-resistant materials. For safe crack growth designs, which are generally associated with high-performance combat aircraft, it is important that the stress corrosion cracks be prevented from occurring or that they be detected before failure, since failure of the parts or components may lead to the loss of the aircraft. As an aircraft ages and protective finishes and coatings break down, concern over part failure caused by SCC becomes more acute. As a result the committee believes that there is a need for the Air Force to periodically assess the susceptibility of their aging aircraft to SCC and take actions to diminish the occurrence of SCC and prevent future part failures. Particular attention should be given to structures that are not designed to be fail-safe.

The committee recommends that the Air Force include an assessment of the vulnerability of each of their aging aircraft to structural failure caused by SCC or SCC combined with fatigue as part of the DADTA updates proposed in this chapter. Specifically, the committee recommends that

- stress-corrosion-critical areas be identified based on past service experience, the susceptibility of the materials to SCC, grain orientations, and probable levels of both applied and residual stresses
- the engineers performing the DADTA update make an evaluation of potential failure modes and consequences of failure for each stress-corrosion-critical area
- protection, inspection, modification, and replacement alternatives be developed as necessary (see recommended short-term research in Chapter 7)

IMPROVED CORROSION CONTROL PROGRAMS

The 1988 accident of the Aloha Airlines 737 aircraft (NTSB, 1988) resulted in much attention being paid to the aging aircraft issue both by the commercial and the military aviation sectors. Although this accident was primarily the result of WFD,[3] it focused attention on all of the factors that can contribute to structural deterioration, including corrosion.

Both the commercial and the military sectors have since taken actions to reduce corrosion and the very high associated maintenance costs.

In the commercial sector, the Air Transport Association and the Aerospace Industries Association in cooperation with the FAA, established the Airworthiness Assurance Task Force to evaluate potential deficiencies in current commercial practices and to provide recommendations and guidance to the FAA and the airline industry on maintaining the structural integrity of 11 different aging aircraft models, including the Boeing 707, 727, 737, and 747; the Airbus A-300; the BAC 1-11; the Fokker F-28; the Lockheed L-1011; and the Douglas DC-8, DC-9, and DC-10. In 1992 the Airworthiness Assurance Task Force was incorporated into the FAA's Aviation Regulation Advisory Committee as the Airworthiness Assurance Working Group (AAWG), shown schematically in Figure 5-2. The AAWG proposed a *mandatory* CPCP to be tailored to each aircraft and operator and implemented by the FAA by airworthiness directives. The need for this program stemmed from fleet surveys, maintenance cost reviews, and comments from operators, all of which pointed to the fact that corrosion resulted in the single largest investment in time and resources in aircraft maintenance programs, and that, in some cases, the aircraft were being maintained in conditions below the manufacturer's expectations. On the other hand, operators that already had comprehensive CPCPs in place experienced much lower amounts of corrosion than those that did not. In fact, if the programs were implemented early in the aircraft's life, the aircraft remained essentially corrosion free. Also, it was noted that operators who utilized liberal applications of corrosion-preventive compounds showed significantly reduced corrosion damage. The essential elements of the AAWG overall CPCP are

- inspection of all primary structures
- initial and repeat inspection intervals based on calendar time rather than flight hours or number of flights
- performance of basic maintenance tasks, including exposure of the corroded area, cleaning, inspection, rework as required, reapplication of corrosion-preventive treatments
- adjustments in the aircraft's overall maintenance program to maintain a corrosion severity of Level I or better (as described below)

The CPCP for each specific type of aircraft was developed by that aircraft's structures task group, which was made up of representatives from the manufacturer, the operators and maintainers, and the FAA.

In the development of CPCPs, the commercial aircraft industry has established severity classification criteria to guide maintenance programs. Corrosion severity is considered to fall into one of the following three classes:

---

[3]Loss of adhesion in the cold-bonded fuselage lap splice contributed to the early fatigue cracking at knife-edged countersunk fastener holes.

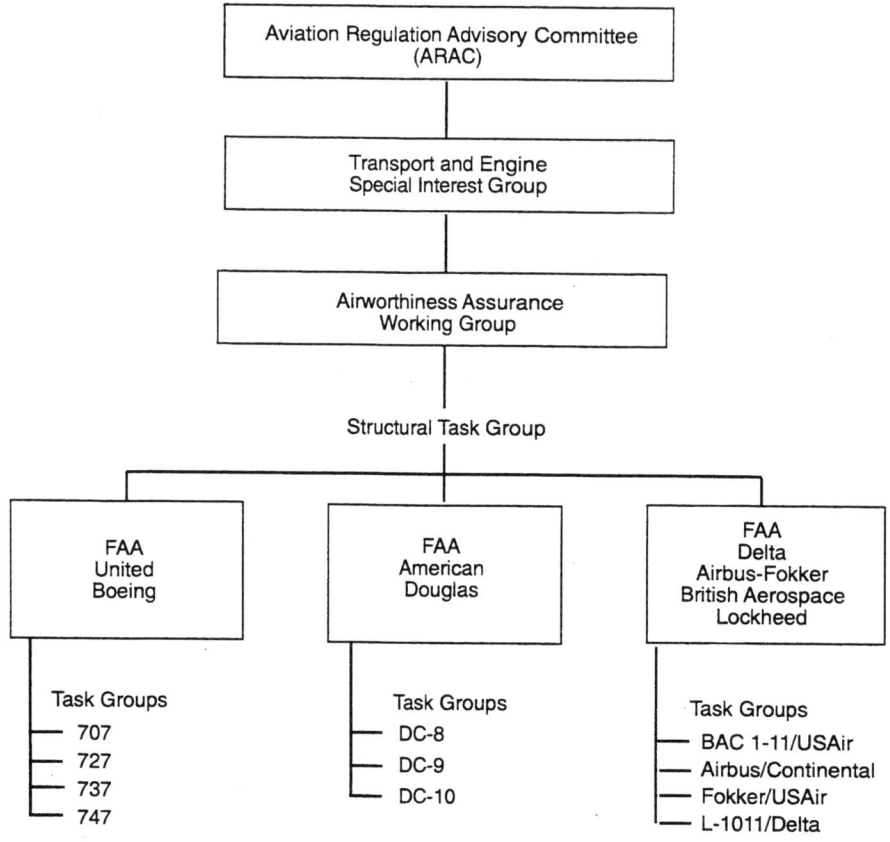

FIGURE 5-2 Organization of commercial aircraft industry aging aircraft working groups. Source: Hidano and Goranson (1995).

Level I corrosion. (1) Corrosion damage occurring between successive inspections that is local and can be re-worked/blended-out within allowable limits as defined by the manufacturer; or (2) corrosion damage occurring between successive inspections that is widespread and can be reworked/blended-out well below allowable limits as defined by the manufacturer; or (3) corrosion damage that exceeds allowable limits and can be attributed to an event not typical of the operator's use of other airplanes in the same fleet (e.g., mercury spill); or (4) operator experience over several years has demonstrated only light corrosion between successive inspections but latest inspection and cumulative blend-out now exceed allowable limit.

Level II corrosion. (1) Corrosion occurring between successive inspections that requires a single re-work/blend-out which exceeds allowable limits, requiring a repair/reinforcement or complete or partial replacement of a principal structural element, as defined by the original equipment manufacturer's structural repair manual, or other structure listed in the baseline program; or (2) corrosion occurring between successive inspections that is widespread and requires a single blend-out approaching allowable rework limits.

Level III corrosion. Corrosion found during the first or subsequent inspections, which is determined (normally by the operator) to be an urgent airworthiness concern requiring expeditious action. Note: When level III corrosion is found, consideration should be given to action required on other airplanes in the operator's fleet. Details of the corrosion findings and planned action(s) should be expeditiously reported to the appropriate regulatory authority (Boeing, 1994:1.1-1–1.1-2).

A CPCP is considered effective if corrosion of identified critical structure is limited to Level I or better.

The intent of these CPCPs is to ensure that corrosion is *never* allowed to progress to the point that it could become a safety issue (hence the emphasis on primary structure). The secondary benefit of the programs is to reduce the operators long-term corrosion maintenance costs.

In the military sector, the Air Force established a Corrosion Program Office at the Warner-Robins Air Logistics Center to oversee and coordinate the Air Force's corrosion prevention and control activities. However, the development, implementation, and execution of specific weapon system corrosion control efforts is the responsibility of the specific system program director. Guidance is provided by Technical Order 1-1-691, which is a tri-service (Navy/Army/Air Force) coordinated manual entitled "Aircraft Weapons Systems Cleaning and Corrosion Control," published January 1992. This manual provides detailed information on such items as preventive maintenance procedures, methods, and materials; inspection techniques; corrosion and paint removal methods and the application of surface treatments; and procedures for applying sealing compounds. Appendix E to this manual is for Air Force use only and contains additional information on aircraft cleaning procedures and intervals as a function of aircraft basing, shot peening and roto peening procedures, and chemical corrosion removal procedures. It is intended that this tri-service manual be used in support of Air Force aircraft manuals and, in the event of conflict, the aircraft manual would take precedence. The Air Force Corrosion Control Office along with the Naval Air Systems Command and the Army Aviation Systems Command are responsible for the maintenance of the manual.

The tri-service manual has a great deal of detailed information on corrosion prevention and control, and a significant effort is being made by the Corrosion Control Office to reduce corrosion in the Air Force's aging aircraft. However, the committee believes that the Air Force does not have the type of comprehensive CPCP for each of its aging aircraft weapon systems on the level of those mandated for commercial airplanes. The committee does not believe that corrosion can or will be completely eliminated in the Air Force's aging aircraft, but with comprehensive programs similar to those established for commercial aircraft, corrosion can be reduced significantly.

The committee recommends that the Air Force undertake the following actions to improve corrosion prevention and control in the aging forces:

- The Air Force's system program directors, in concert with the appropriate major commands and the Corrosion Control Office, should perform an internal audit of each of the Air Force's commercial-derivative aging aircraft (i.e., the E-3, E-8, E-4, V.C-25, T-43, C-137, C-18, C-22, KC-10, and C-9) to ensure that the corrosion control programs are in full compliance with the CPCPs mandated for commercial counterparts. In addition to the primary structures covered by the commercial programs, the Air Force should ensure that adequate corrosion control measures are being applied to corrosion-susceptible secondary structures.
- The Air Force's system program directors, in concert with the appropriate major command and the Corrosion Control Office, should review the detailed corrosion control programs of each of the Air Force's aging aircraft listed in Table 3-1 that is not scheduled to be retired in the near future (i.e., the KC-135, C-5, A-10, B-52H, B-1B, F-15, F-16, C-130E/H, U-2, and T-38) and upgrade them as necessary to a level equivalent to or better than the CPCPs that are mandated for commercial aircraft. Again, corrosion-susceptible secondary structures as well as the primary structures should be included in the programs.
- The Air Force's ALCs, with the Corrosion Control Office, should evaluate the applicability and cost effectiveness of dehumidification, as described in Chapter 4, to reduce the likelihood of corrosion.

ECONOMIC SERVICE LIFE ESTIMATION

As discussed in Chapter 4, major economic impacts can be expected to occur with the onset of WFD in fail-safe-designed aircraft structures and with the rapid growth in the number of fatigue-critical areas in safe-crack-growth-designed aircraft structures. When either of these occur, the options are to modify the structure, replace major portions or components of the airframe, or retire the aircraft. If the economic impact is sufficient to justify retirement, this would constitute the economic service life of the aircraft. However, there are a number of other factors that also contribute to the economic service life, and this should be viewed from the broader perspective of the total cost to operate an aircraft system. There are several examples in which it has been cost effective to modify or replace major components of an airframe, even when they have experienced WFD. Some of these aircraft have continued in service for many more years (e.g., the KC-135, C-5A, and C-141). On the other hand, it appears quite possible that the economic burden of operating a given type of aircraft could become excessive before the onset of WFD or the rapid rise in fatigue-critical areas. For example, it was pointed out in Chapter 4 that corrosion (including SCC) is currently the most costly maintenance problem for the Air Force's aging aircraft. If not substantially diminished in the future through improved prevention and mitigation measures, corrosion damage, either by itself or in combination with fatigue cracking, could cause the Air Force to undertake major modifications, major component replacements, or perhaps aircraft retirement.

Clearly, as was pointed out in Chapter 2, there is a need for an overall economic service life estimation model that integrates the estimates of structural deterioration caused by fatigue, corrosion, and SCC with all other operating cost elements. The current lack of such a tool inhibits Air Force planners from establishing a realistic time table to phase out a current system and to begin planning for replacement aircraft. Some examples of cost elements that should be tracked and projected for inclusion in such a model are related to

- field-level personnel, facilities, materials
- depot-level personnel, facilities, materials
- acquisition and repair of repairable parts
- acquisition and repair of consumable parts
- support equipment
- field-level sortie generation: fuel, maintenance production
- depot maintenance program: programmed depot maintenance, analytical condition inspection, and speedline production
- structural and subsystem modifications: repair and maintenance technology insertion, safety, mission capability
- field-level maintenance: isochronal inspections
- engine depot overhaul program
- sustaining engineering
- environmental impacts

In addition to these cost elements, there are several operational metrics that can be used by the aircraft system program managers to develop an overall assessment of a system's operational effectiveness, such as

- mission capability rate
- sortie generation/abort rate
- "not mission capable" rates
- maintenance man hours per flight hour
- depot flow time and quantity of aircraft in depot status
- parts cannibalization rate
- accident rate

The problem of service life estimation is complicated not only by the technical difficulties involved in predicting the onset of WFD and the growth in fatigue-critical areas and the numerous factors affecting structural deterioration caused by corrosion, but also by the interrelationships and the relative importance of the many cost and operational metrics listed above. Ideally, the service life estimation model should utilize the best possible technical estimates of the major structural modification and/or component replacement times, account for the cost and operational metrics listed above, and balance and weigh their relative importance.

The committee recommends that the Air Force make a concerted effort to develop a credible service life estimation model or methodology that would be accepted by the Air Force senior management and the Department of Defense decision makers (e.g., the Defense Acquisition Board) as the authoritative guide for supporting replacement decisions and budget inputs. Such an analysis could be considered to be analogous to the cost and operational effectiveness analysis (COEA) that is undertaken early in a weapon system acquisition cycle to support milestone decisions, but in this case would be done later in the system life cycle to support a modification/update or replacement decision. When the model is completed, it is recommended that it be used to update the service life estimates for the Air Force's aging aircraft listed in Table 5-1.

CONTINUED ENFORCEMENT OF THE AIRCRAFT STRUCTURAL INTEGRITY PROGRAM

The Air Force has been very successful in controlling structural fatigue failures for more than two decades. One of the primary factors contributing to this success has been the rigid enforcement of the ASIP. Internal compliance by Air Force management was directed by Air Force Regulation AFR 80-13, and contractor compliance was achieved by making MIL-STD-1530 and supporting specifications part of the weapon system contract. Placing ASIP on contract ensured that the damage-tolerance-based inspection and maintenance requirements would be developed, and the AFR ensured that the Air Force would follow through with their implementation, including the incorporation of adjustments to the inspection and modification times brought about by changes in aircraft use. ASIP has also provided industry with guidance on all of the design, analysis, and test requirements necessary to achieve the aircraft's design service life goal and has provided Air Force engineers the basis for making sound technical recommendations to system program directors concerning the aircraft structure. It is for these reasons that the committee is very concerned that ASIP, per MIL-STD-1530 and its supporting specifications, will no longer be placed on aircraft acquisition and modification contracts due to former-Secretary of Defense Perry's initiative to reduce the use of government specifications in acquisition programs.

The guidance provided under the initiative directs that the intent of rescinded specifications be incorporated, if appropriate, into contracts through performance requirements, thereby giving contractors wider latitude and greater discretion in how to meet them. The Air Force is presently converting AFR 80-13 to an Air Force Instruction and the ASIP standards and specifications to a "guidance document" for use by government and industry for executing the program. Although these are important first steps, the committee does not believe that they go far enough. The committee believes that the "guidance document" approach will still be vulnerable to inconsistent interpretation and application between the various program offices within the Air Force. This approach will also leave industry uncertain as to the acceptability of their ASIP-related engineering practices to the various government weapon system program offices and result in ASIP provisions in a program that is more vulnerable to programmatic cost reductions. The end result can be incomplete or omitted ASIP tasks that would seriously degrade the effectiveness of the FSMP, which is designed to protect the structural safety of the aircraft.

Short of reinstating AFR 80-13, MIL-STD-1530, and supporting specifications, the committee recommends that the

Air Force take the lead in pursuing the development of a National Aerospace Standard for ASIP. Such a standard would result from the coordinated efforts of the military services and industry as to what constitutes an acceptable and affordable ASIP for new aircraft acquisitions and modifications to existing systems. The standard would be issued by industry and referenced by the government as a measure of acceptable compliance with contractual ASIP performance requirements. This approach would effectively communicate the government's requirements to industry and reduce the likelihood of inconsistent application and execution of ASIP tasks. It is anticipated that the end result would be a continued high level of operational safety and improved force structure management.

TECHNICAL OVERSIGHT AND RETENTION OF TECHNICAL CAPABILITIES

Much of the success of the Air Force ASIP during the past two decades can also be attributed to the competency of the ASIP managers and the engineering support groups within the maintenance organizations and the technical oversight provided by an Air Force Matériel Command, Aeronautical Systems Center (AFMC/ASC) standing committee that has guided the many DADTAs that have been performed. In addition, various Air Force Scientific Advisory Board (SAB) and Division Advisory Group ad hoc committees have contributed to this success.

The aging aircraft engineering disciplines that have been developed and typically reside within the ALC's technology and industrial support engineering (TIE) organizations include specialists in nondestructive inspection, stress analysis, design of structural repairs, fracture mechanics analysis, failure analysis, and corrosion control. However, these groups also draw on the expertise of AFMC/ASC engineering and the Wright Laboratories for assistance on specific problems. Where major modifications or a detailed knowledge of the aircraft design are involved, the ALC normally contracts with the original equipment manufacturer for the required assistance, which has been the case for most of the DADTAs that have been performed over the years.

The committee believes that, in recent years, the Air Force's capability to support ASIP and perform structural assessments has deteriorated somewhat as a result of budget and manpower reductions and grade-level limitations within the ALCs. Unfortunately the reduction in capability comes at a time when the need for capabilities has been increasing because of the aging of the force. ASIP managers are burdened with day-to-day maintenance problems and program cost and schedule pressures that allow them little time to focus on the broader issues such as implementing improved corrosion controls or obtaining improved estimates of when to expect the onset of WFD. Also, there seems to be considerable variability in the engineering capabilities among the different ALCs, perhaps because of insufficient policy direction and oversight from AFMC headquarters (HQ AFMC). Finally, there is no single technical focal point to coordinate ASIP, the supporting DADTAs, and the aging aircraft structures issues. A standing committee that at one time monitored DADTAs has been discontinued.

Although the committee *does not* believe that these apparent reductions in technical capabilities and oversight are currently jeopardizing structural safety, the prognosis for the future of the aging force is not optimistic unless the following near-term actions are taken.

First, the committee recommends that HQ AFMC form an aging aircraft engineering resources group consisting of engineering management representatives from AFMC headquarters, Aeronautical Systems Center's engineering and technical management organization (ASC/EN), Wright Laboratories, and each of the ALC TIE organizations. This group should be chartered to examine the quantity and quality of the engineering skills in each of the aging aircraft disciplines that are available at the ALCs, ASC/EN, and Air Force Laboratories and to compare these skills with the projected requirements over the next five years. Where imbalances exist between skill requirements and skill availability, the group should examine alternate methods of fixing the imbalances (e.g., redistribution of available resources, hiring contract engineers, more contracted assistance from the original equipment manufacturers, proposed changes in grade structure, or proposed additional military and civilian positions) and prepare a recommended course of action for Air Force senior management.

Second, the committee recommends that an aging aircraft technical steering group (AATSG) be formed that reports to the commander of AFMC and whose chair is a member of the Air Force SAB. This group should meet no less than two times a year, but can meet more often if so desired by the commander of AFMC. The method of operation would be similar to the existing division advisory groups, which implies that the chair report to the SAB steering committee on a semiannual basis. The purpose of the steering group would be to monitor and provide guidance to the various recommended near-term engineering and near- and long-term research activities discussed in this report and to report on progress and, as necessary, potential problems. They would also provide advice and surveillance over near- and long-term research programs to ensure seamless transition of technologies (6.1 through 6.7) into aging aircraft. The members of the AATSG would be selected by the Air Force, in consultation with the SAB, from the government, industry, and academia and represent the various aging aircraft technical disciplines.

Third, the committee recommends that five technical working groups be formed (i.e., one for each of the five basic elements of the proposed near-term and long-term R&D programs as shown in Figure II-2). These working groups would consist of technical specialists from the Air Force Office of Scientific Research, Wright Laboratories, ASC/EN, and the ALC system management and TIE organizations and would form the technical link from basic research (6.1) through implementation (6.7). These groups would be responsible for understanding and interpreting user needs and ensuring that the R&D efforts in each of the five basic elements are focused on meeting these needs in a timely and economical manner.

Finally, the committee recommends that HQ AFMC appoint a single knowledgeable and experienced technical leader responsible for the oversight of the aging aircraft engineering and the near-term and long-term R&D activities recommended in this report. The selected individual would serve as the primary point of contact with the AATSG and the internal technical working groups and would have the authority to provide the overall day-to-day technical direction to the structural aging aircraft program. The selected individual should report to the appropriate management level within AFMC so as to be given the authority and stature necessary to execute the assigned tasks.

TECHNOLOGY TRANSITION INTO AGING AIRCRAFT

One of the most effective ways for increasing the reliability and speed of nondestructive evaluation and reducing the costs of repairing aircraft with structural cracking and corrosion problems is through the transition of improved technologies into application. In the past, this has been difficult for the system program directors because the links with technology development activities (e.g., labs, industry, other services) were not well established. As a result, system program directors often acquired technologies to solve their specific weapon system problems using internal sustaining-engineering funds. Often, these initiatives required a modest amount of development and in many cases had generic characteristics that would permit application to other systems. Very seldom, however, were these technologies made available to, or embraced by, other system program directors. In addition, existing technologies available in industry or in other services often went undiscovered. Considerable improvement has occurred in recent years as a result of the AFMC technology master process described in Chapter 2, which created a linkage between the technology users (system program directors, ALCs, major commands) and the technology producers (laboratories, industry, other military services). There is clear evidence of substantial improvement in the number of laboratory technology programs that focus on the problem of aging aircraft. Although this progress is evident in programs involving 6.1, 6.2, and 6.3 funding, there has not been a commensurate improvement in the programs that implement technology into aging aircraft (e.g., 6.4, 6.5, 6.6, 6.7). System program directors still rely primarily on producibility, reliability, availability, maintainability, sustaining engineering, and manufacturing technology funding to bring emerging technologies to bear on aging aircraft problems. These funding categories are typically funded well below requirements, some are limited to one year for expenditure, and some have limited application and low funding thresholds per individual project. The solution to this problem is to provide seamless funding of aging aircraft technology transition programs from 6.1 through 6.7.

The committee believes that the concept of a seamless funding–budgeting link from 6.1 through 6.7 for aging aircraft initiatives is very attractive. It is based on the implicit assumption that the project is fully prepared for implementation at the next level. For this to be the case, the principal investigators must fully understand the requirements at the next level, and at the same time they must exercise enough discipline in conducting the study to ensure that the project is able to make the transition at the earliest possible time. This will foster teamwork between the technology developer and the technology users. A considerable effort will be required to make the transitions as straightforward as possible. The initial step, developing a clear definition of the problem and the results required, is of key importance. The important point is for the technology-developer and the technology-user communities to approach aging aircraft technology problems as an integrated team.

The committee recommends that 6.1, 6.2, and 6.3 aging aircraft technology programs that are generic and have potential for wide application not be approved through the technology master process unless it is linked to an appropriate 6.4 through 6.7 program to provide transition to force application. It is critical to the success of the aging aircraft program that a seamless funding–budgeting link be created from development through application. Furthermore, the five technical working groups recommended by the committee should be responsible for ensuring that there is a seamless link in funding for the program from 6.1 through 6.7.

6

Research Recommendations: Fatigue

LOW-CYCLE FATIGUE

As described in Chapter 4, there are two primary technical issues related to low-cycle fatigue:

- the rapid increase in the number of fatigue-critical areas in safe-crack-growth-designed structures and the potential for missing new areas as they develop
- the onset of widespread fatigue damage in fail-safe-designed structures

Currently, the primary method for identifying fatigue-critical areas is through a detailed examination of the locations where cracking occurs during full-scale fatigue testing of the aircraft. These findings are supplemented by data from stress analyses, strain surveys, and experience with similar design details, materials, and material forms that have been prone to cracking on other aircraft. Occasionally, fatigue-critical areas that were not previously identified are found during tear-down inspection of actual force aircraft (e.g., during maintenance or inspections of high-time aircraft). However, for safe-crack-growth-designed aircraft, reliance on in-service inspections to identify new critical areas can be extremely dangerous. To be assured that accidents will be avoided, cracks must be found before reaching critical size. For some aircraft structures, these critical sizes can be very small.

Despite the committee's efforts to develop a research initiative that would improve on the current approach for identifying new fatigue-critical areas, no viable near-term or long-term research activities were identified. Likewise, the current Air Force research program has no ongoing or planned research in this area. The committee can only emphasize the extreme importance of using all available full-scale test and service experience data and state-of-the-art stress analysis methods to perform the durability and damage tolerance assessments (DADTAs) recommended in Chapter 5 so that all fatigue-critical areas can be identified. This is particularly important for the high-priority DADTAs (i.e., for the F-16, A-10, T-38, and U-2), all of which concern aircraft that are of non-fail-safe designs. Currently available finite element and solid-modeling stress analysis techniques should be considered for those cases in which the structures have not been analyzed using these modern methods. Fatigue test articles that have not been evaluated in detailed tear-down inspections should be evaluated (if test articles are available). If necessary, additional fatigue testing or detailed tear down of high-time aircraft should be performed. This is the most critical task in the DADTA for non-fail-safe structures.

Air Force research projects in low-cycle fatigue focus on widespread fatigue damage (WFD), specifically on the development and validation of analysis tools to predict the onset of WFD and on corrosion–fatigue interactions. Program plans for WFD include

- basic research tasks that include efforts to investigate (1) analysis methods for multiple-site damage, (2) formation of cracks from manufacturing and service-induced defects, and (3) three-dimensional nonlinear fracture predictions; also included is a new initiative that includes fundamental research to characterize and analyze WFD
- applied research to (1) develop analysis methods to model the effects of WFD, (2) determine initial quality for use in risk analysis, (3) evaluate the effect of WFD on crack growth, (4) upgrade the *Damage Tolerance Handbook*, (5) develop in-service and experimental WFD data, and (6) develop process sciences methodology for metallic structures
- exploratory research to perform a structures demonstration for WFD

The committee believes that the Air Force program in WFD, as originally presented, indicated an incomplete understanding, among at least some of the researchers, as to the nature and failure scenarios associated with WFD (NRC, 1997). The concern arose from discussions of plans to evaluate the remaining life of structures with WFD. As discussed in Chapter 4, the onset of WFD is the safety limit, beyond which the aircraft should not fly without modification or replacement of the structure. Consequently, remaining life is not an issue once the structure is in the state of WFD. Recent revisions to the Air Force R&D program address the committee's concern. The committee has identified several particular strengths in the planned research in WFD:

- the program emphasis on configurations applicable to military aircraft (e.g., thick wing structure and integrally stiffened structures)

- the stated intention to experimentally verify fail-safe residual strength prediction methodology with large components or panels
- the effort to determine the initial quality of typical structure for use in structural life and risk analyses, which is an essential element in the prediction of when small widespread fatigue cracks will exist in service aircraft
- basic research focused on the formation, growth, and distribution of small fatigue cracks from small manufacturing or service-induced defects and corrosion damage
- the development of advanced probabilistic methods for force risk assessment
- the work in the area of processing science that could lead to higher-quality materials and tighter process controls that can increase resistance to fatigue crack initiation
- the effort to update the Air Force's *Damage Tolerance Handbook*
- coordination with Federal Aviation Administration (FAA) and National Aeronautics and Space Administration (NASA) research on WFD to ensure that their efforts are complementary

The committee believes that the current engineering approach to WFD should be supplemented with advanced analysis methods and more extensive use of the results of detailed tear-down examinations of full-scale fatigue test articles and retired aircraft.

Near-Term Research and Development

Recommendation 1. Extend and validate recent advances in nonlinear finite element modeling and fracture mechanics to the unique configurations of fail-safe-designed military aircraft for the prediction of residual strength.

Although some emphasis has been placed on the prediction of the fail-safe residual strength of military aircraft structures (i.e., thick wing structure), the committee suggests a critical review of current methods used to determine the fail-safe residual strength levels for the many different detailed structural configurations that exist in military aircraft that could be prone to WFD. Typical configurations of interest include large pressure doors and door hinges, ramps and ramp attachments, canopy attachments, wing-to-fuselage and fin-to-fuselage attachments, multiple adjacent fuselage frames, circumferential fuselage joints and chordwise wing tension joints, chordwise wing splice joints, and engine attachment structures. Where improved methods appear to be necessary, they should then be developed and experimentally verified.

Most engineering fracture mechanics methods assume linear elastic behavior. A number of investigations have established that the use of elastic-plastic fracture mechanics is essential to determining the residual strength of an airframe structure with WFD (Harris et al., 1995; Atluri, 1997). Improved methods are required to treat the effects of plasticity on the fatigue crack growth and fracture behavior typically exhibited by the ductile alloys used in aircraft construction. The committee recommends that the Air Force evaluate ductile fracture criteria for three-dimensional crack configurations and integrate the criteria into analysis methods to predict residual strength. The research should consider the effects of alloy composition, material product forms, structural configurations (e.g., thick, heavily loaded components), and exposure to aircraft environmental conditions.

Recommendation 2. Improve current methods to determine the onset of WFD by (a) comparison of full-scale test articles with tear-down inspection of service aircraft components and metallurgical examinations of full-scale fatigue test articles and (b) critical examination of the procedure for extrapolating the sample of cracks documented during a tear-down examination to generate a distribution function that may be used in a risk assessment.

Although the committee endorses the longer-term R&D efforts to develop analytical methods to predict the initiation and growth of cracks to the sizes at onset of WFD, the primary method to determine the onset of WFD in the near term will be estimates based on empirical data (e.g., full-scale fatigue test results or, if available, tear-down inspection results from operational aircraft), combined with fail-safe residual strength analyses. Because tear-down inspection of actual fleet aircraft entails the destruction of one or more aircraft (or major portions of aircraft) and comes too late to provide data for force planning, the Air Force has been primarily dependent on the results of full-scale fatigue testing to assess WFD. Unfortunately, full-scale fatigue test results are not necessarily representative of the actual operational load spectrum and generally neglect the potential influence that environmental exposure may have on the crack initiation process.

There is no defined effort in the Air Force research program to improve the current method of estimating the onset of WFD in the aging aircraft program. The committee recommends research to assess the validity of (and if necessary, suggest improvements to) the approach to estimation of onset of WFD.

Long-Term Research and Development

Recommendation 3. Conduct experimental research to establish the relationship between the physical basis for crack formation/nucleation and crack distribution functions.

The development of analytical prediction methods for crack initiation, based on rigorous descriptions of initiation processes, would be extremely complex to develop because of several mechanisms and the wide variations in conditions that may be involved at any given structural location. The committee does not believe that rigorous analytical models can be developed that accurately consider all of the various mechanisms and conditions involved in fatigue crack initiation.

Nevertheless, small crack theory uses the equivalent initial flaw (EIF) approach with initial flaw sizes determined from microstructural features characterized by microscopy rather than back calculating from fatigue data (Ritchie and Lankford, 1986; AGARD, 1990). Although predictions of total fatigue life of laboratory test specimens using fracture mechanics analysis methods and initial crack sizes determined from microstructural features have been shown to be accurate, microstructural defects are only one of several possible root causes of fatigue crack initiation. Therefore, the committee believes that the most promising analytical approach to predict the behavior of other initiating mechanisms is to use an EIF size determined from experimental data. A comprehensive EIF-based fracture mechanics approach, including simulative experimental methods for the prediction of initiation and growth of small cracks, is vital to the development of analytical prediction capability for the onset of WFD. The committee suggests the development of an EIF database, correlated with full-scale structural test articles, for cracks that initiate because of fretting, very small defects, scratches, dings, and corrosion damage.

Recommendation 4. Develop and experimentally verify analytical methodology to predict crack distribution functions.

The quantification of the principal parameters—aircraft use spectra, initial quality, stress level, and structural geometry—needed to provide analytical estimates of the time- and use-dependent crack populations and the associated fatigue life and critical crack sizes requires an extension of the existing analytical methods and approaches. The committee suggests that the most promising approach is to combine existing deterministic tools for the prediction of stress levels, residual strength, and crack growth with existing risk analysis tools to account for statistical variability of the situations that might lead to failure of the aircraft. However, to deal effectively with the problems associated with variations of initial quality, local construction, stress level, and use spectra, it is necessary to have an integrated hierarchical approach that uses structural analysis and risk management methods.

The range of cyclic loading conditions that contribute to the development of fatigue cracking may result in crack populations that are unique to each aircraft type and structural location. Therefore, the analytic representation of crack population as a function of service time is extremely difficult because the crack population depends on events and conditions that can only be quantified either in a worst case deterministic sense or bounded in a statistical sense. Because of these inherent uncertainties in developing a unique crack population for each aircraft, probabilistic risk assessment methods are necessary adjuncts to deterministic methods. As described in Chapter 4, current risk assessment analyses use data obtained from aircraft component tear-down examinations to account for the uncertainties in estimates of fatigue crack characteristics and distribution.

Recommendation 5. Validate analytical methods using results of laboratory and full-scale fatigue tests, tear-down inspections of structural components removed from retired aircraft, and experimental tests of built-up structure.

The results of numerous full-scale fatigue tests and tear-down examinations of structural components removed from retired aircraft are already available as a benchmark for validating the advanced analysis methodology. The data obtained from the near-term research Recommendation 2 should be used to the extent possible. Additional carefully defined critical tests with well-characterized boundary conditions and loading histories will also be required to fully verify all aspects of the analysis methodology. The methodology should be verified by comparison with test data obtained from several different aircraft structural components and loading conditions that are susceptible to WFD.

HIGH-CYCLE FATIGUE

The Air Force aging aircraft program related to high-cycle fatigue is included in the research plan for structural dynamics. The dynamics program includes research tasks in predictive methods and suppression techniques. Topics include

- acoustics and sonic fatigue
- structural dynamics
- computational methods
- health monitoring
- structural repair and component replacement

The most important elements of the current dynamics research program are upgrade of the design guide for aft body and airframe aeroacoustics and acoustic fatigue, design and test of new structural repairs and components, buffet load alleviation, unsteady aerodynamics and aeroelastic codes, and health monitoring (where it is related to dynamics load definition and temperature and chemical environment definition).

The committee believes that the program would be improved if emphasis were placed on dynamic loading and high-cycle fatigue degradation specifically associated with aging of in-service aircraft. Much of the current Air Force program, as described to the committee, contains technology development programs that are generally related to the design and analysis of emerging aircraft systems and not to the life extension of existing systems. These programs probably are needed, especially those that are basic research and technology development, but funding should be separated from the aging aircraft budget.

The committee recommends that near-term and long-term research focus on dynamic loading cases that are related specifically to aging aircraft. Suggested near-term research opportunities include efforts to improve methods to determine dynamic response. Recommended long-term research extends the near-term program to include characterization of threshold crack growth behavior, analytical prediction of dynamic response, expert systems for the design and analysis of repairs, and dynamic load monitoring and alleviation.

Near-Term Research and Development

Recommendation 6. Improve and verify methods to predict dynamic strains and deflection responses of unrepaired and repaired structures and improve laboratory and flight test methods for measuring structural response.

The response of structure under critical load conditions must be determined before a repair of a dynamically loaded structure can be performed successfully. This is crucial because the repair must be sufficiently durable to provide structural integrity under these loading conditions, and because the dynamic response of the structure is often affected by the repair. This change in response is key to developing a long-lasting repair that does not induce further damage in surrounding structures.

It is often difficult to determine the dynamic response for structures subjected simultaneously to high- and low-cycle fatigue loads. In many cases, both loading conditions must be included to accurately simulate the failure and to develop a long-lasting repair. Improved ground testing or flight testing methods to determine the structural response under dynamic load conditions will be important in verifying the driving forces and structural responses responsible for early cracking or cracking in aging structures.

Long-Term Research and Development

Recommendation 7. Characterize threshold crack growth behavior for materials and structures used in Air Force aircraft. Examples of specific tasks include

- determination of the relationship between conventional fatigue endurance limits and crack growth threshold stress intensity factors

- evaluation of the sensitivity of crack growth thresholds to aggressive environments, such as humidity, saltwater, fuel, or hydraulic fluids
- modification of current test methods or development of new low-cost methods to develop crack growth threshold stress intensity factors
- estimation of fatigue life under high-cycle and mixed high-cycle–low-cycle regimes for intact and repaired structural components

Dynamic fatigue failures are very sensitive to the threshold crack growth rates of the materials involved, which are related to the time to initiate cracks from inherent defects within the material or to the surface finish or roughness of the finished part (Bucci et al., 1996). Maintaining sufficient fatigue life in the presence of dynamic loading requires either maintaining very low vibratory stress levels or increasing controls on material defects and design details (stress concentrations) that can lead to early fatigue failures. Knowing the relationship between conventional crack initiation behavior and threshold crack growth for the materials of interest in the Air Force aging aircraft could be valuable in the development of low-cost methods to determine the effects of high-frequency loads on fatigue.

Threshold crack growth behavior, and therefore dynamic fatigue life, is very sensitive to the effects of aircraft environments, including humidity, saltwater, fuel, or hydraulic fluids. These environmental conditions can reduce loads at the threshold crack growth regimes by as much as a factor of two and dynamic fatigue life by as much as an order of magnitude. The committee believes that the environmental sensitivity of dynamic fatigue behavior must be determined, validated by test, and documented specifically for materials used by the Air Force.

Generally, threshold crack growth test methods involve shaker table testing of sheet materials. These methods generate large numbers of cycles in very little time, but the crack growth data are not usually measured (Beier, 1997a). Modification of current tests to correlate with threshold data or development of cost-effective methods that characterize the relationship between threshold crack growth rates and time to crack initiation for uncracked samples are needed. The potential for linking naturally occurring flaw growth to a corresponding threshold crack growth rate would be a significant outcome of this development.

Life prediction for dynamically loaded structures is difficult to achieve with accuracy, given the sensitivity of the life to the threshold load levels. Also, the combined loading at both low- and high-cycle frequencies complicate this prediction problem notably. Some recent strides have been made in life prediction under combined high- and low-cycle fatigue loading (Saff and Ferman, 1986). But the problem remains one of determining the root source of the problem and modeling this root cause properly. The committee recommends an effort to validate life predictions under both dynamic and combined high- and low-cycle fatigue to provide Air Force maintenance organizations with the capability to rapidly and accurately predict the lifetimes of both structures and repairs subjected to dynamic loading conditions.

The problem of life prediction of the repaired structure is similar to that of the unrepaired structure, except that the effect of the repair on load paths, mode shapes, and loading frequencies must be predicted accurately before the life prediction can be accurate. Given the uncertainty in the original life predictions noted above, the committee recommends that a program be performed to determine the potentially significant effect of repairs on component fatigue life.

Recommendation 8. Develop and validate through laboratory or flight tests analytical methods to predict dynamic response of aging structures and repairs. Include consideration of affected structure away from the repair/modification and the accelerating effects of environmental exposure.

Methods to analytically predict structural dynamic response, validated through the laboratory or flight tests, are required to assess aging structures and repairs. In some cases, structural repairs to address high-cycle fatigue serve to exacerbate the dynamic loading problems because local repairs of dynamically loaded structures can move the failure to a new location defined by the repair itself. Analytical methods must be capable of determining the response of the structure beyond the repairs.

Recommendation 9. Develop and implement an expert system, based on analytical methods and previous experience, to aid the design and analysis of repairs or modifications (both damped and undamped) of components susceptible to high-cycle fatigue damage. Examples of specific tasks include

- development by the original equipment manufacturer of a database of dynamic loading conditions for particular locations on the structure and the acceptable frequencies and duration of the response in those locations
- determination of damping levels for the repaired structure required to achieve the desired frequency range for the structure and the damped repair configuration (stand-off damping or adhesive layer or stiffening)

Given the potential and the capabilities afforded by today's materials, the ability to apply damped repairs should be pursued. These repair systems have been studied for two to three years and are nearing the point at which flight demonstration is becoming feasible (Beier, 1997b; Rogers et al., 1997). Nevertheless, there are repair considerations for dynamically loaded structures, for both damped and undamped repairs, to ensure successful repairs. These special considerations include

- critical modes and responses (and natural frequencies) of the original structure
- critical modes and responses of the repaired structure
- response level required to obtain the desired life of the structure
- driving force behind the cracking that was the root cause of the original problem

The complexity of the analysis of dynamically loaded structures and repairs, often further complicated by high static or low-frequency loads, make these structures and repairs excellent candidates for the development of an "expert system." This system would have the data required from the original equipment manufacturer imbedded within the system to define the primary modes and responses of dynamically critical structures, or those structures known to have given trouble in a particular airframe. It would have the capability to design conventional or damped repairs and would be capable of assessing the durability of both the structure and the repair under the loads known to be in that portion of the structure. Such systems are becoming more user friendly as software and hardware capabilities improve. Experimental systems are being evaluated by the Air Force laboratories and the air logistics centers (Rogers et al., 1997). The key to these expert systems for repair of dynamic structures is the successful prediction of the environment and the response of the repaired structure.

Recommendation 10. Develop improved dynamic load monitoring and alleviation technologies that take advantage of recent advances in sensors and controls and computational capabilities. Examples of specific opportunities include

- improved load and condition-monitoring capabilities using piezoelectric sensors and neural networks for data analysis
- active flutter suppression and buffet load suppression systems that link condition-monitoring capabilities described above with piezoelectric transducers/actuators and intelligent controls technology

Dynamic loads lend themselves to relatively easy detection and measurement in flight. Simple accelerometers and strain gages can be applied for dynamic load tracking in the same way that maneuver loads are being recorded in several fatigue tracking systems. These systems were originally implemented in the 1960s and 1970s when the on-board computational capabilities were limited. However, with the enhanced speed and memory of today's computers, both dynamic and maneuver loads can be measured directly and recorded for postflight analysis.

Newer smart structures technologies such as piezoelectrics and neural networks are available that enable improved load/health monitoring as well as alleviation of dynamic loads

(Geng et al., 1994; Kim and Stubbs, 1995). Neural networks provide the potential to monitor more locations on the aircraft while reducing the number of sensors required. Piezoelectric-based health monitoring systems have been demonstrated in the laboratory for integrated damage detection of both metallic and composite structures (Lichtenwalner et al., 1997).

Intelligent control systems have been developed and demonstrated to suppress flutter and buffet loads using both conventional control surface actuators and piezoelectric actuators. Piezoelectric transducers alternatively can sense dynamic response and input dynamic loads that can be used to counteract the external loading conditions. The application of these sensor/actuators to the suppression of dynamic loads has been demonstrated in the laboratory for scaled aircraft models.

These technologies should be transitioned to full-scale structures and, assuming successful results, demonstrated under flight conditions in order to prepare them for implementation in Air Force aircraft to detect and react to dynamic loads. The primary research effort is to determine computationally efficient methods of handling and interpreting large amounts of data and storing only what is needed to make the status of the structure clear. Along with the proper mix of sensors (e.g., accelerometers; pressure transducers; or piezoelectric sensors, actuators, or strain gages) to best determine the environment and response, the system, at best, must be capable of rapidly assessing damage location and the extent of damage, systems affected, and severity of the damage when interrogated on the ground or in the air.

CORROSION/ENVIRONMENTAL EFFECTS

As described in Chapter 4, the committee is concerned that, as structures age, as corrosion protection systems continue to deteriorate, and as materials corrode, there may be effects that have not been adequately considered. Specific corrosion concerns or issues that could affect safety limits and inspection intervals for safe-crack-growth-designed aircraft and the onset of WFD in fail-safe aircraft include

- the influence of corrosion on applied stresses resulting from material thinning and local bulging or pillowing of thin sheet due to buildup of corrosion products
- the potential influence of corrosion on material mechanical properties (i.e., toughness, strength, elongation) resulting, for example, from the absorption of hydrogen by the metal during the corrosion process
- the potential influence of corrosion and corrosive environments on crack growth rates below the threshold for stress corrosion cracking

In the current Air Force program, corrosion and environmental effects on fatigue are part of a category of projects labeled "corrosion–fatigue." Also included in this category are the research and development efforts in corrosion prevention and control, which are discussed in Chapter 7. The principal topic areas in the Air Force program that relate to corrosion and environmental effects on fatigue are:

- analysis of corrosion effects on structural durability
- test protocol development for corrosion–fatigue interactions
- analysis method demonstration and validation

The committee has pointed out several strengths of the Air Force program, including fundamental efforts to characterize and analyze corrosion and the potential effects of corrosion damage on fatigue behavior, an effort to update the Air Force *Damage Tolerance Handbook* to include corrosion effects, and efforts to coordinate with FAA and NASA research on corrosion to ensure that efforts are complementary (NRC, 1997). However, the committee believes that the Air Force program overemphasized characterization, evaluation, and prediction of corrosion effects and had insufficient emphasis on prevention and control technologies, particularly from a materials and processing perspective. This large emphasis on the effects of corrosion on structural durability may be in response to the recommendations of the Materials Degradation Panel of the 1994 summer study of the Scientific Advisory Board, which had a similar emphasis (SAB, 1996). Although the committee recognizes the need for some specific research activity in this area, the primary focus should be on the development and institutionalization of corrosion prevention and control as discussed in Chapter 7.

The committee believes that it is important for the near-term program to address the specific concerns that have been expressed concerning the procedures that the Air Force and industry use to account for corrosion and environmental effects on fatigue-crack-growth-based safety limits and inspection intervals for safe crack growth structures. Specifically, the concern is about the potential influence of corrosion and environment on the growth of cracks from the assumed manufacturing flaw size (typically 0.05 in.) to either the critical size or the threshold size for stress corrosion cracking. Also, the committee believes that it is important to assess the potential influence that the induced bending stresses from corrosion-caused pillowing has on the fail-safe residual strength of fail-safe-designed structures. The influence of corrosion and environmental exposure on the initiation and growth of the very small cracks associated with the onset of WFD is a less urgent need in the near term. This is because the current basis for predicting the onset of WFD is the result of tear-down inspections of actual high-time operational aircraft, which have been exposed to the real operational environment. In some cases, these components contain severe corrosion (e.g., see the discussion of the E-8 fuselage panel tear-down inspection in Appendix A).

As noted above, it is a long-term goal to be able to analytically predict the onset of WFD based on the initiation and growth of very small fatigue cracks. To achieve this goal, the committee believes that there is a need for fundamental research to provide a basic understanding of corrosion and environmental effects of fatigue crack initiation and growth to sizes associated with the onset of WFD (i.e., as small as 0.04 in.). It is anticipated that this fundamental understanding will also contribute directly to the development of improved corrosion prevention and control procedures.

Near-Term Research and Development

Recommendation 11. Determine if prior corrosion damage has an effect on basic material properties such as modulus, yield strength, and fracture toughness.

It has been suggested that long-term aging in a corrosive environment may also result in changes to basic material properties such as modulus, percent elongation, yield strength, and fracture toughness. The committee suggests that corroded components removed from retired aircraft be evaluated in an experimental study to determine if long-term material aging in a corrosive environment produces changes to basic material properties. The goal of this work is to definitively lay to rest the issues of whether corrosion damage affects fundamental material properties or if property loss attributable to corrosion is related only to loss of material. If an effect is detected, it must be quantified with respect to the effect on design allowables.

Recommendation 12. Determine potential effects of prior corrosion or exposure to a corrosive environment on fatigue crack growth.

Typically, current practice is to develop crack growth rate data (i.e., da/dn data) for use in safety limit calculations in a wet or humid environment. Crack growth rates, in addition to being sensitive to exposure environment, can be dependent to some degree on frequency, particularly at low stress intensities (i.e., very small crack sizes) and at stress intensities above the threshold for stress corrosion cracking. Past assessments have not considered the effect to be significant in the determination of safety limits and inspection intervals for the materials, crack sizes, and stress levels typically involved in combat aircraft. However, this issue should be revisited. Specifically, this work should determine if prior corrosion affects the fatigue crack growth rates over the range of crack sizes (and stress intensity values) typically associated with the determination of safety limits (e.g., from 0.05 in. to critical size). Tests should be conducted for typical alloys (e.g., for 7075 and 2024 aluminum plate), in both wet and dry environments, and for a minimum of two cyclic frequencies and two stress levels (R values). Also, material thinning from corrosion will result in an increase of the stress level, which in turn will increase the crack growth rates. This can be accounted for easily in the determination of safety limits and inspection intervals by assuming a specific amount of allowable thinning. This is currently being done on some of the older aircraft (e.g., the KC-135).

Recommendation 13. Assess the effect of widespread corrosion-caused pillowing on the fail-safe residual strengths of thin-skinned fuselage splice joints.

It is accepted that WFD will severely degrade the fail-safe residual strength of fuselage structure (e.g., the residual strength in the presence of a two-bay crack). It is also known that extensive pillowing or bulging of fuselage lap splices has occurred in some aircraft fuselages as a result of corrosion products in the splices in the absence of WFD (e.g., in some E-8 aircraft). The concern is that the high induced stresses caused by pillowing could potentially degrade the fail-safety of the fuselages prior to the onset of WFD. The committee recommends that an experimental research effort involving the fail-safe testing of one or more large panels that contain pillowed splice joints be defined and executed to resolve this issue.

Long-Term Research and Development

Recommendation 14. Perform fundamental research to determine if there are unique material or environmental conditions that promote the growth of small fatigue cracks under typical aircraft loading conditions. Examples of specific tasks include

- evaluation of the effects of various levels of prior corrosion and environmental spectra (i.e., chemistry, temperature, mechanical variables) on the development and growth of cracks to sizes typical of the onset of WFD in representative aircraft structure
- modification of existing high-humidity tests or development of improved accelerated testing protocols to simulate corrosion-fatigue interactions representative of severe aircraft service

The normal testing environment for corrosion fatigue of aircraft aluminum alloys is humid air at 25°C. Although this is an aggressive environment for aluminum alloys, it may not represent the worse case scenario for fatigue during flight. The initiation and propagation of small fatigue cracks that lead to WFD generally occur in areas that are occluded (e.g., lap splice joints, fastener holes, etc.). Prior corrosion associated with the land-based environment will most likely affect the actual flight environmental spectra (chemistry, temperature, mechanical variables) in the critical areas associated with WFD. The committee recommends that representative environmental spectra be determined and used to evaluate environmental effects on the development and growth of cracks from approximately 25 microns to sizes typical of the onset of WFD in representative aircraft materials and structures. The focus of the research should be to determine if there are unique material or environmental conditions that promote the growth of small fatigue cracks under typical aircraft loading conditions.

Recommendation 15. Perform fundamental research to determine how the nature of an existing flaw (i.e., flaw morphologies, pits, intergranular cracks, machine defects) in conjunction with severe environmental conditions (developed above) affects fatigue crack growth from very small cracks to the size associated with WFD.

The geometry and location of the cracks, as well as whether they are transgranular or intergranular, may have an effect on the local chemistry and thus on the fatigue growth rates. The committee recommends that research efforts be undertaken to determine how the nature of an existing flaw (i.e., flaw morphologies, pits, intergranular cracks, machine defects) in conjunction with the worst possible environmental spectra, affects fatigue crack growth from very small cracks to the size associated with WFD.

Recommendation 16. Perform fundamental research to determine the extent to which hydrogen governs the growth of small fatigue cracks relevant to the onset of WFD, as well as high cycle fatigue crack growth. Examples of specific tasks include

- determination of the effect of local, dissolved hydrogen in fatigue crack growth from small cracks
- assessment of local hydrogen content as a common indicator for the prediction of the effect of corrosion on subsequent fatigue behavior

During corrosion processes of aluminum and its alloys, hydrogen is normally dissolved. The amount of dissolved hydrogen depends on the chemistry of the environment (for example, more hydrogen is dissolved when NaCl is present than in a normal high-humidity atmosphere), the temperature, the chemical potential, the alloy, and the temper (Leidheiser and Das, 1975; Smith and Scully, 1996). Hydrogen is known to have an adverse effect on fatigue resistance. For example, it has been shown in laboratory tests of a high-purity Al-Zn-Mg alloy that preexposure to humid air causes reductions in fatigue resistance that are comparable to those resulting from exposure to water vapor during fatigue testing, an effect that was completely reversible by vacuum storage timed to permit hydrogen diffusion out of the samples (Ricker and Duquette, 1988). The committee recommends fundamental research to determine the extent to which hydrogen governs the growth of small fatigue cracks relevant to the onset of WFD, as well as high-cycle fatigue crack growth. The goal of the recommended research is to determine if local, dissolved hydrogen participates in and exacerbates fatigue crack growth from small cracks for the alloys and tempers pertinent to aging aircraft and if local hydrogen content can provide a common indicator for predicting the effect of corrosion on subsequent fatigue behavior.

7

Research Recommendations: Corrosion and Stress Corrosion Cracking

CORROSION PREVENTION AND CONTROL

The economic burden that corrosion presents to the Air Force has been reported widely as the single most expensive structural maintenance issue, affecting both operating costs and readiness. The Air Force Scientific Advisory Board Materials Degradation Panel cited estimates of the costs associated with corrosion-related detection and repair range from $1 billion to $3 billion annually (SAB, 1996). The ideal solution is to prevent corrosion from starting. However, complete corrosion prevention should be considered a research challenge because, despite prevention efforts, corrosion will continue to occur in Air Force aging aircraft. Therefore, the committee recommendations reflect the reality of anticipating and controlling corrosion problems ranging from barely detectable to widespread. In addition, the recommended research reflects the need for immediate engineering solutions to get the aircraft out of the depots quickly, as well as long-term research so that future operations can practice effective control and prevention.

Research efforts in corrosion prevention and control are currently part of the corrosion–fatigue category of the Air Force aging aircraft program. The current Air Force research relating to characterization of corrosion–fatigue interactions is discussed in Chapter 6. In addition, the program on corrosion–fatigue includes some effort on evaluation and characterization of improved corrosion-resistant materials. The research program includes

- basic research involving characterization and analysis of corrosion, fatigue damage development, and environmental and corrosion effects; also included is a new initiative to investigate the development of pitting corrosion in aluminum alloys
- applied research to develop in-service and experimental corrosion and fatigue data and efforts involving evaluation and characterization of improved corrosion-resistant materials and corrosion chemistry
- exploratory research that investigates fine-grain processing to improve corrosion resistance

The committee supports efforts to improve the definition of corrosion damage metrics and the associated test protocols, to characterize in-service corrosion damage to provide data for severity assessments, and to characterize and analyze corrosion. However, the current program emphasizes characterization and evaluation over prevention and control technologies and does not provide maintenance handbook-level guidance to upgrade corrosion resistance of operating forces through alloy substitution and application of materials and processing advances.

The suggested Air Force research in corrosion places much more emphasis on early detection of corrosion and implementation of effective corrosion control and mitigation practices. In general, the committee recommends short-term program emphasis on corrosion detection and maintenance technology (i.e., how to deal with existing corrosion) and longer-term emphasis on the fundamental understanding of corrosion and characterization of corrosion rates and the development and institutionalization of corrosion prevention and control practices. The committee believes that a practicable and more cost-efficient strategy for dealing with corrosion damage of airframe structures is needed to effectively guide prevention, control, and force management decisions for aging aircraft. The research topics emphasized in this approach include improved protective coatings, advances in alloys and processes offering improved corrosion protection, improved techniques to discover and quantify hidden corrosion without requiring disassembly of the aircraft (see Chapter 8), and methods to predict corrosion rates to guide inspection intervals and repair/modification activities. These developments along with the implementation of improved corrosion prevention and control actions described in Chapter 5 (including classification of corrosion severity, expanded use of corrosion-preventive compounds, and, potentially, dehumidified storage) will prevent physical corrosion from progressing to a point where it would limit the structural life of Air Force aircraft.

Near-Term Research and Development

Recommendation 17. Establish the link between service environment and laboratory test conditions and develop a laboratory test protocol to perform accelerated testing that more accurately simulates corrosion damage experienced in aircraft service.

Current accelerated aging practices strive to reproduce three service conditions in laboratory tests:

- corrosion type (e.g., pitting, intergranular corrosion, etc.)
- damage severity (e.g., depth of attack)
- corrosion product chemistry

Although these methods provide comparisons between materials, they do not adequately simulate corrosion processes and rates that occur in service.

The committee recommends that the Air Force take a somewhat different approach by developing methods that simulate the damage in a quantifiable manner. This approach builds on standard accelerated test practices, but adds the quantitative aspect that can be used to develop damage metrics and provide a link to corrosion reaction kinetics in service environments. Significant effort within R&D laboratories in the development of test methodologies must be guided by field data that at a minimum define the relevant ions, humidity cycles, temperature cycles, and UV radiation intensity.

Recommendation 18. Evaluate the durability of new environmentally compatible protective coatings. Examples of specific tasks related to aging aircraft include

- characterization of the role of stress, both static and cyclic, as a source of initial defects in coatings
- evaluation of coating durability in a fretting environment and in a crevice corrosion environment
- evaluation of the effects of chemical and physical heterogeneity within coatings on the long-term performance
- characterization of the effects of new paint removal techniques such as sodium bicarbonate, wheat starch blasting, and pulsed cold plasmas on corrosion resistance and the performance of subsequently applied coatings
- determination of the effects of thermal and physical aging on the adhesion characteristics of primer coats and conversion coatings

The Air Force has long recognized that the durability of protective finish systems is the most important factor, other than resistance to discrete mechanical damage, in the development of corrosion for aging aircraft (Miller, 1987). Aircraft coatings must meet a demanding set of criteria, including (1) ambient curing, (2) long-term corrosion protection and adhesion to a wide variety of substrates, (3) resistance to environmental chemical exposure (e.g., hydraulic fluids, fuels, solvents, and cleaning solutions), (4) long-term exterior durability with minimal change in optical or physical properties (Hegedus et al., 1995), and (5) mechanical durability to operating stresses and in fretting environments.

The epoxy and polyurethane systems that have been the mainstay of aircraft coatings have been modified and will continue to change in response to environmental regulations that limit the release of volatile organic compounds (VOCs) and heavy-metal-containing materials such as chromium or cadmium used to inhibit corrosion (NRC, 1996a). Candidate technologies to reduce these releases include water-borne and high-solids coatings to reduce VOC release and nonchromate additives including molybdates, nitrates, borates, silicates, and phosphates (Hegedus et al., 1995). In general, these technologies have failed to exhibit the corrosion protection and durability of conventional systems. Recognizing these concerns, the Air Force has a research program to develop and validate environmentally compatible coatings. The aging aircraft program needs to assess the durability of these coatings under simulated service conditions using the accelerated testing protocol in Recommendation 17.

An area of critical need is the development of effective coating removal and surface preparation methods (AGARD, 1992). Surface blasting with wheat starch and sodium bicarbonate has been shown to be effective but not without several drawbacks (i.e., paint removal rates are slow and nonuniform, very large quantities of blast materials are needed, and residual surface contamination remains following cleaning). The residual surface material is suspect in diminishing the performance of subsequently applied coatings. New paint removal methods must be examined. One promising technique utilizes a pulsed cold plasma that has the capability of converting paints to the gaseous state for safe collection. The plasma energy can be controlled very sensitively so that sublayers can be removed should it be desired to leave the primer intact. In addition, the plasma is not a line-of-sight method and can therefore remove paint from within crevices. This technology is demonstrable at this time and could be developed into a useable prototype within two to three years.

Evaluation of improved materials and processes should take into account the complex interactions present in a real system, particularly between the different surface finish layers, and the materials compatibility and durability issues. The goals of the research are to (1) rate new corrosion-preventive compounds (CPCs) and protective coatings, (2) assess aging effects caused by thermal and environmental exposure on the adhesion characteristics of replacement primer coats and conversion coatings, and (3) qualify environmentally compatible protective coatings for Air Force use.

Recommendation 19. Evaluate and implement methods to provide earlier detection of corrosion. Examples of specific tasks include

- investigation of environmental sensors to allow aircraft maintenance organizations to anticipate when conditions are likely to lead to corrosion
- evaluation of the applicability of the Navy's condition-based maintenance program to Air Force needs

- development of techniques to locate, monitor, and characterize defects and chemical and physical heterogeneity within coatings

Corrosion control programs rely on the early identification of corrosion before significant material loss occurs (Agarwala et al., 1995). Corrosion detection can be accomplished using nondestructive evaluation (NDE) inspections (see Chapter 8) or health monitoring technologies. Prototype corrosion microsensors that detect currents associated with galvanic corrosion have been demonstrated by the Navy (Agarwala and Fabiszewski, 1994). The sensors are thin enough to be applied to corrosion-prone and hidden areas. The sensors have been applied successfully in the laboratory to evaluate the integrity of coatings, sealants, hidden structures, and organic composites. Field trials are under way with the ultimate goal of using the sensors to provide data for the Navy's attempt to implement condition-based maintenance of corrosion-prone structure (Moore, 1997). The committee recommends that the Air Force investigate selective application of corrosion-sensing technologies and validate promising techniques under service conditions.

Long-Term Research and Development

Recommendation 20. Initiate a basic research effort to support the development of improved materials and methods for corrosion prevention and control. Examples of specific tasks include

- identification and generalization of the mechanisms by which coatings (particularly chromates), CPCs, paints, and adhesives provide protection; in particular, provide information about the interaction of organic paint systems with the aluminum surface oxide
- identification of deterministic factors in corrosion pit initiation and localized coating breakdown

A range of both ongoing and new R&D opportunities exist for the prevention and mitigation of corrosion in aircraft structures. Many of the drivers for the development of new coatings and coating processes are the impending Environmental Protection Agency and Occupational Safety and Health Administration mandates to eliminate silica from surface cleaning methods, chromates from conversion coatings and primers, and VOCs from cleaning solutions and paint compositions (AGARD, 1996). Thus, two simultaneous objectives must be met in that new technologies must be identified that are both environmentally acceptable and effective as corrosion mitigators. A blue ribbon panel has recommended that the corrosion protection mechanisms provided by chromates used as conversion coatings and as corrosion-inhibiting pigments be established. A multi-university research initiative, headed by Ohio State University, has been established to explore corrosion protection mechanisms of chromate primers.

Boeing has devoted extensive R&D to remove chromates from the conversion coatings and primers. The cobamine process is an effective alternative to chromate conversion coatings; however, it is unclear whether the heavy-metal content (in the form of cobalt) of this process chemistry will be acceptable from a waste water standpoint. Synergistic combinations of rare earth compounds have proved effective as corrosion-inhibiting additives for primers with bulk solution studies; however, they must now be tested in primer coatings. It cannot be assumed that once incorporated into a polymer matrix, a compound will be an effective corrosion inhibitor. Boeing is also studying sol gels as a combined replacement for the conversion coating and primer (Blohowiak et al., 1997). Although these coatings have excellent adhesion, it is becoming apparent that the incorporation of corrosion-inhibiting additives will most likely be needed to bolster the corrosion protection properties of sol gels.

Other chromate replacement chemistries, such as alkaline oxide baths, are being examined. Conventional carbonate chemistries have proved effective for the non-copper-bearing alloys, but have had limited success for the 7XXX and 2XXX alloys. Very recent modifications of these bath chemistries have created promising corrosion-protective films on 7075-T6 and 2024-T3 (Buchheit, in press). Other variants of the hydrotalcite coating process are being explored as a possible means to achieve low-contact-resistance surfaces and active corrosion protection (Taylor et al., 1997).

Recent exploratory investigations suggest that coatings of quasicrystalline materials, applied using environmentally benign processes, could provide corrosion resistance (Dubois et al., 1993). The applicability to aging aircraft and performance in an aircraft environment has not been investigated.

Recommendation 21. Characterize corrosion rates for the major types of corrosion. Examples of individual tasks include

- quantification of the influence of environmental and materials variables, including inhibitors, on corrosion rates
- development of analytical models of corrosion initiation and growth to provide quantitative information to support repair–replace decisions

The characterization of corrosion rates for the major types of corrosion identified in Chapter 4, including uniform or general corrosion, galvanic corrosion, pitting corrosion, fretting corrosion, crevice (filiform and faying surface) corrosion, intergranular (including exfoliation) corrosion, and stress corrosion cracking, will provide valuable information to aircraft operators to support repair–replace decisions and to establish inspection and maintenance intervals. In addition, a quantitative understanding of corrosion rates will help to establish requirements for sensitivity and reliability in the development and validation of improved NDE methods as recommended in Chapter 8.

Recommendation 22. Conduct basic research to determine the fundamental factors that govern coating durability. Examples of specific tasks include

- determination, for example, using localized electrochemical and chemical measurement techniques, of the effect of exterior environmental chemistry (including gases), coating resin chemistry, and the substrate surface chemistry on factors that lead to stable growth of coating defects
- investigation of adhesion mechanisms between coatings and relevant substrate materials to determine the role of coating adhesion in the long-term performance of coatings on metal substrates
- investigation of environmental effects on surface chemistry and morphology of new conversion coatings and subsequent adhesion of organic coatings
- development of analytical models to predict long-term coating performance based on materials and interfacial characterization following short-term exposures

STRESS CORROSION CRACKING

As described in Chapter 4, stress corrosion cracking (SCC) is an environmentally induced, sustained-stress cracking mechanism associated with exposed short-transverse end grain in thick plate, extrusions, and forgings made from susceptible alloys. SCC is driven predominately by residual tensile stresses remaining from material heat treatment or fit-up, but can also be triggered by operational loads and forces from the buildup of corrosion by-products. The best SCC defense is prevention, rather than controlling its growth. The committee suggests that the near-term research program of the aging aircraft program focus on developing data and documenting results that would lead to affordable upgrades in SCC prevention and component repair and modification procedures. The recommended focus of the long-term R&D is on establishing fundamental materials and microstructural effects on SCC susceptibility and a basic scientific understanding of SCC mechanisms to support efforts in prevention.

Near-Term Research and Development

Recommendation 23. Develop data and document results that would lead to affordable upgrades in SCC prevention. Examples of specific tasks include

- development of resource guide(s) and databases that catalog significant items; fleet survey results; best practices and common problems; and SCC ratings of the various materials, manufacturing processes, protective systems, corroding environments, and repair practices
- development of cost information and tools in easily accessible form for analyzing various SCC prevention and repair options
- development of an alloy substitution matrix to allow for the replacement of susceptible alloys with improved materials

Appreciable time has passed since the original design of many older Air Force models. In the intervening years, significant advances have been made in alloys, protective systems, and in the understanding and control of grain flow and residual stress in thick wrought products, most notably forgings and extrusions. Although SCC resistant materials (e.g., 7050, 7150, 7055) and tempers (e.g., T73, T74, T76, and T77 tempers for 7XXX-series Al alloys and T8 tempers for 2XXX-series Al alloys) are now available, high-susceptibility materials (namely, 7075, 7079, and 7178-T6 and 2024-T3) remain in wide use, particularly in the older models. A life extension program would presumably aim to maintain component performance similar to the original. However, to avoid future SCC problems, replacement parts could be made from materials with improved resistance to SCC, particularly if plans for the retrofit include verification testing of components. Likewise, SCC-resistant tempers of steel alloys could be considered in similar fashion.

The assessment of materials and manufacturing processing interchangeability is necessary to take advantage of materials and process advances. This work would support substitutions that would decrease susceptibility of older aircraft to SCC in aging aircraft and to support repair–replace decisions. Currently, Air Force operators manage SCC separately for each aircraft with alloy substitution generally addressed on a part-by-part basis. The development of a common data and experience base that described vulnerable structures, susceptible alloys, protection and repair processes, and assessments of costs would reduce redundant engineering efforts and lead to guidelines and a justification for modification efforts to improve SCC resistance (Bucci and Warren, 1997). For example, some alloys could be considered as generally equivalent with predecessor alloys (e.g., 7050 for 7079), whereas others could be considered preferred replacements within specified limits (e.g., 7150 or 7055-T7X for 7075-T6).

Recommendation 24. Perform a systematic evaluation of the sensitivities and effectiveness of various protective systems on prevention and control of SCC.

The recommended work is particularly concerned with the evaluation of the effect of measures taken to prevent fatigue and corrosion on SCC susceptibility. Examples of fatigue and corrosion prevention measures include prestressing techniques such as cold working, peening, and laser shock processing (Ratwani, 1996); material effects; surface finishes; CPCs (water displacing); corrosion-inhibiting elastomeric sealants; bonded doubler repairs (with and without reinforcing fibers); organic coatings; and inorganic corrosion-protective systems. The key issue is to determine if such measures adversely affect SCC resistance. The results of these investigations will support decisions on how to maintain structures to reduce susceptibility to corrosion, fatigue crack growth, and SCC.

Recommendation 25. Conduct research to better understand the cause and effect of manufacturing and assembly stresses, the variability of these residual stresses within and among categories of components, and potential routes for their alleviation.

An important consideration in avoiding SCC, although often neglected, is the effects of fabrication (e.g., mill working history, heat treatment, machining, straightening and forming, and fit-up stresses) introduced during part manufacture and assembly. Higher residual stresses from manufacturing operations were determined to be a major factor in the SCC problem for thicker parts. Tensile stresses in the short-transverse direction relative to the metal grain structure, rather than stresses imposed by service loads, were found to be by far the most frequent driving force for SCC. In such cases the direction and magnitude of the tensile stresses are not typically recognized and accounted for during the design process.

The committee believes that research to better understand the cause and effect of these stresses, their variability within and among categories of components, and potential routes for their alleviation—particularly for thick, complex parts—would support efforts to document SCC vulnerability and to develop SCC protection alternatives. Also important to the anticipation of SCC is understanding and controlling the impact of metallurgical grain flow in the completed part, including the effect of prior process history.

Examples of specific tasks include

- assessment of the impact of residual stress and grain flow on past and potential future SCC problems to categorize as either singular events or symptomatic of a much greater problem within the aging fleet
- investigation of the means to minimize tensile and residual stresses, which may become significant during fabrication and assembly
- quantification of the potential degrading influence of intrinsic residual stresses on benefits from peening, cold work, coatings, and other protective systems

Recommendation 26. Quantitatively evaluate the SCC susceptibility of current Air Force materials (alloy and product forms), based on experimental SCC threshold stress data, fracture mechanics threshold stress intensity data, and crack growth kinetics.

Although testing of actual structural components returned from service is generally preferred, this often is not practical because of limited sample availability and the size and complexity of service components. Hence, laboratory-scale methods are needed to evaluate the SCC susceptibility of alloys used in Air Force aircraft. Among the many test methods available for evaluation of SCC, two basic approaches have emerged. One approach is based on pass-fail testing of smooth or unintentionally flawed specimens to determine a threshold stress below which SCC will not occur. The second approach is based on fracture mechanics testing of specimens with intentional cracks to determine both the threshold stress intensity factor (K_{ISCC}) and the kinetics of crack growth (i.e., da/dt). Both approaches are used by industry and government laboratories to evaluate SCC susceptibility of materials (Spowls et al., 1984). The results of this work will demonstrate the validity of uniting SCC initiation and propagation test and evaluation approaches and will support the long-term research task (Recommendation 27) to develop improved criteria to rate material and system SCC performance in a way that is consistent with the current structural integrity methods (Bucci et al., 1986).

Long-Term Research and Development

Recommendation 27. To support the recommended emphasis on SCC prevention and control, conduct fundamental research in the following areas

- mechanisms that drive SCC and experimental determination of SCC kinetics
- small crack mechanics and the associated test and probabilistic methods
- role of material and component/assembly manufacturing processes parameters to define their interchangeability potential (e.g., replacement of forgings with machined plate that involves the effects of texture, grain flow and residual stress)
- characterization of process/microstructure/performance relationships (i.e., grain structure, residual stress) and development of models to describe and predict SCC behavior
- development of improved evaluation criteria to rate material and system SCC performance in a way that is consistent with structural integrity methods
- development of predictive models for residual stress and stress relaxation processes

There has been a large body of research in SCC that have been useful in identifying components that are susceptible to SCC. The principal gaps in previous research are in the areas of (1) SCC prevention methods for old materials that explicitly consider processing/form/microstructure/performance relationships and (2) determination of consequences of SCC over time. The committee recommends fundamental research to improve basic and scientific understanding of microstructure, process, and performance linkages and how they scale from laboratory to full-scale structure. Systematic, fundamental work is needed to define and develop practicable, predictive tools (e.g., model(s), input data, validation testing, and design criteria) based on a sound understanding of the underlying physics, mechanics, metallurgy, and design and manufacturing processes.

Recommendation 28. Develop models and methodology for life prediction for structures susceptible to SCC.

Currently, SCC life prediction is limited because there are no workable computational models of SCC processes. The field is plagued with confusion created to a large extent by (1) the complex, multifaceted nature of the phenomenon, which involves metallurgy, mechanics, chemistry, and kinetics; (2) the large number of variables known to affect SCC behavior; (3) relatively poor correlation between laboratory test results and service experience; (4) extensive data scatter; (5) difficulty in assessing precisely the service conditions that a part must withstand; and (6) unknown internal stress states (e.g., residual stress).

8

Research Recommendations: Nondestructive Evaluation and Maintenance Technology

The management of an aging aircraft force relies on airframe inspection and maintenance and repair programs to ensure that the inherent safety and reliability imparted by the structural design are sustained, deterioration is detected, and, when deterioration occurs, structural integrity is restored. Effective maintenance of airframe structure requires nondestructive evaluation technology capable of reliably detecting all flaws larger than the maximum allowable size, structural evaluation and assessment tools to support repair–replace decisions based on inspection results, guidelines for preventive maintenance, and design and processing methods for structural modification and repair. This chapter presents near-term and long-term research in nondestructive evaluation and maintenance and repair that support the development of an integrated approach to life-cycle management of aging aircraft.

NONDESTRUCTIVE EVALUATION

Nondestructive evaluation (NDE) is a pivotal technology in the management of the aging fleet. If the NDE technology is effective and applied in a timely fashion, efficient inspections and management decisions can be made to either return aircraft to service or to assign them to modification or repair. Such decisions depend on the reliability of NDE inspection capabilities and can significantly affect either safety or economics if made incorrectly. The development of NDE technology for aging airframe structures is driven by structural requirements and cost considerations. Proper application of NDE technology can offer significant improvements in diagnostic capabilities and provide characterization of damage to direct structural repair requirements. In addition, NDE methods must be able to detect all flaws larger than the maximum allowable size and introduce quantifiable and direct characterization of structure and material condition.

The Air Force sponsors broad NDE efforts spanning basic research (6.1) administered by the Air Force Office of Scientific Research focused on new NDE technology, applied (6.2), and exploratory (6.3) research administered by Wright Laboratories Materials Directorate and several engineering development or evaluation study programs administered by the air logistics centers (ALCs), most notably Oklahoma City ALC efforts on NDE for corrosion detection. Elements of these generic and technology-based programs are currently being realigned and refocused to address the needs of the aging aircraft program. The Air Force R&D program on NDE for aging aircraft is focused on two primary topics: (1) corrosion detection and characterization and (2) detection of cracks, including sizes associated with the onset of WFD.

The corrosion detection and characterization category includes

- basic research to investigate and demonstrate innovative NDE techniques that have the potential to produce significantly improved accuracy of defect detection and characterization and reliability for detection of corrosion and small fatigue cracks
- applied research to evaluate various NDE approaches for corrosion detection, including neutron radiography, optical fiber sensors, and neutron activation analysis; ribbon x-ray sensors; x-ray spectroscopy; nonlinear electromagnetic methods; and enhanced methods to detect incipient corrosion
- an exploratory research project to conduct depot-level demonstrations of successful methods from applied research efforts, evaluate data fusion and image analysis methods for NDE data evaluation, and demonstrate and validate high-resolution real-time radioscopy systems
- a limited manufacturing research effort to evaluate NDE methods for corrosion in aging airframes

The widespread fatigue damage (WFD) and fatigue crack detection category includes

- basic research (as described above) that includes both corrosion and small fatigue crack detection techniques
- exploratory research in improved methods and equipment for NDE of supersonic turbine engines, small crack detection methods, remote sensing of fatigue, and hidden flaw detection
- short-term projects to develop inspection methods and determine inspection reliability for multilayer crack detection for the C-141 and to develop prototype thermography inspection systems and procedures to inspect composite and bonded structures

The committee has obtained data, which are included in Appendix A, on the structural problems experienced in many of the Air Force's aging aircraft. Based on these service experiences it is apparent that there are both specific and overarching features to the aging aircraft NDE needs. Specific needs (described in Chapter 4) include the development of techniques to detect (1) fatigue cracks under fasteners, (2) small cracks associated with WFD, (3) hidden corrosion, (4) cracks and corrosion in multilayer structures, and (5) stress corrosion cracking in thick sections. As pointed out in Chapter 4, current NDE methods qualified for a given aircraft application for the detection and characterization of a particular flaw will not necessarily be applicable directly to another application, even though detectability requirements are the same, because of variations in geometries and materials. Consequently, even though only two phenomenon-based flaws are listed (cracks and corrosion), the actual NDE engineering problem base is many times larger.

A majority of the current Air Force NDE research effort relating to aging aircraft is aimed at the discovery of techniques to detect and characterize fatigue cracks and corrosion. Significant efforts in these topics are also funded by the FAA (FAA, 1996) and NASA (Winfree, 1996) as well as smaller efforts by other agencies and by industry (SPIE, 1996). The committee encourages continuing in-depth interactions between the Air Force and efforts supported by others to enhance the overall impact of the Air Force efforts. However, the current program appears to put relatively little emphasis on the development of new tools that will enhance the cost effectiveness of NDE systems, including system design and development, validation, and force-wide application. Although the research programs indicate that their efforts are coordinated with the ALC needs (e.g., the application of ultrasonic creep wave techniques to the C-141 weep hole cracking problem), programs for the general field validation and implementation of technology developed in the research program have been inadequate. This inadequacy should be addressed by the recommended improvements in linkages from technology development through implementation, which are discussed in Chapter 5. Formal validation and demonstration arrangements similar to those used by the FAA (i.e., the Aging Aircraft Nondestructive Inspection Validation Center; Walter, 1995) should be considered.

The committee recommends that the Air Force pursue a two-pronged R&D effort to develop inspection technology for aging aircraft. In the near term, efforts should be directed at improving the Air Force NDE technology base by evaluating, validating, and implementing currently available NDE technology to address key aging aircraft problems. In addition, the near-term program should explore and apply new engineering approaches to develop quantitative NDE inspections that are much faster, less costly, and that result in a technology base that is more flexible and easily managed in treating the diversity of aging aircraft problems. In the long-term program, the committee believes that the current empirical approach to validation of new NDE methods should be augmented with analytic approaches to develop reliable, quantitative NDE methods. Emphasis should be placed on NDE technique design and development aimed at improved detection reliability and defect characterization, cost-effective validation and qualification procedures, transferability to a range of applications, and interdisciplinary coordination with other elements of the aging aircraft strategy.

Near-Term Research and Development

Recommendation 29. Evaluate, validate, and implement currently available NDE equipment and methods for use at Air Force maintenance facilities to address specific aging aircraft problems. Focus near-term efforts on inspection capabilities needed to support the inspection requirements resulting from the DADTA updates that are recommended in Chapter 5.

Inspection of aging aircraft requires an integrated NDE approach to effectively address critical inspection needs identified in Chapter 4. Efforts should be initiated to evaluate, adapt, and utilize NDE advances and know-how developed by Air Force programs, other federally funded programs, and commercially developed technology for detection of corrosion and cracks.

In addition to the Air Force efforts, a number of specific advances have been made in NDE during the past several years that may provide solutions for some of the aging aircraft needs. Examples of advances that have been realized in a number of methods include eddy currents (Wincheski et al., 1994, 1997; Moulder et al., 1995, 1996; Bieber et al., 1997), ultrasonics (Hsu and Patton, 1993; Komsky et al., 1995; Komsky and Achenbach, 1996; Barnard and Hsu, 1997), thermal wave imaging (Emeric and Winfree, 1995; Favro et al., 1995, 1996; Syed et al., 1995), radiographic methods (Ting et al., 1993), magneto-optic methods (Fitzpatrick et al., 1996; Thome et al., 1996), and quantum interference devices (SQUIDS) (Ma and Wikswo, 1996; Podney and Moulder, 1997). The preceding advances, although not generally commercially available, have had some degree of evaluation with industry or at the FAA Aging Aircraft Validation Center. Some benchmarks for comparison of these advances with off-the-shelf methods are available. For example, for eddy current methods, it has been shown that the standard state-of-the-art practice for detectability of cracks under fasteners is about 0.10 in. long and that the best achievable using laboratory equipment is 0.040 in. long under aluminum and 0.050 in. long under steel (Spencer and Schurman, 1995; Hagemaier and Kach, 1997). Recent reports indicate that the ultrasonic "dripless bubbler" (Hsu and Patton, 1993) and pulsed thermal wave (Favro et al., 1996) techniques had been successful in detection of corrosion in various configurations (Howard and Mitchell, 1997).

Critical inspection needs, examples of candidate techniques, and suggested validation applications for aging aircraft are presented in Table 8-1. The list of potential techniques in Table 8-1 should not be considered comprehensive; other candidate techniques could be applicable. Recommended applications for validation efforts are based on the committee's current knowledge of potential problem areas as discussed in Appendix A. Additional specific applications can be expected to be defined as a result of the recommended durability and damage tolerance assessment updates.

The committee recommends that the Air Force use a life-cycle approach to evaluate and validate methods that

TABLE 8-1 Critical NDE Inspection Needs for Aging Aircraft

Critical Need	Candidate NDE Methods	Potential Techniques	Potential Validation Aircraft
Fatigue cracks under fasteners	Electromagnetic	Magneto-optic imaging Pulsed eddy current Eddy current arrays	B-1, F-15
	Thermal	Time-resolved thermography	
	Ultrasonic	Laser ultrasonics Scanning UT probes EMAT transducers	
Small cracks associated with WFD	Ultrasonic	Guided waves EMATs Laser ultrasonics	E-8, VC-137, C-18
	Electromagnetic	Scanning pulsed eddy current Microwave imaging (60–90 GHz) Large-area magneto-optic	
	Thermal	Time resolved, scanning	
Hidden corrosion	Electromagnetic	Pulsed eddy current	KC-135, A-10, C-5, C-130
		Multifrequency eddy current SQUID technology, eddy current	
	Thermal	Time-resolved thermography	
	Radiography	Energy-sensitive detectors Microfocus real-time radiography Neutron	
	Ultrasonic	Bubbler/scanning methods	
	Optical	Boroscope	
Cracks or corrosion in multilayer structures	Electromagnetic	Pulsed eddy current Multifrequency eddy current	KC-135, A-10, C-5, C-130
	Radiography	Real-time imaging In-motion film	
	Ultrasonic	Scanning (if gaps can be bridged)	
Stress corrosion cracking in thick sections	Ultrasonic	Pulse echo, scanning	C-5

consider detectability and inspectability, full-scale validation, material degradation mechanisms, technique reliability, inspection intervals, and cost. It is important to implement a structured selection and implementation regimen that includes

- down-selection of candidate methods based on damage characterization and performance requirements
- validation of the down-selection including development of probability of detection (POD), detectability functions, sizing considerations, orientation, and location of the defects or damage areas
- implementation of the NDE method inclusive of operating limits, equipment performance, and test procedures

The process used to down-select methods is considered to be the critical step in the evaluation of candidate methods and should include a clear assessment matrix tabulation of the potential candidate NDE methods versus weighted requirements. Such methodology (e.g., Cepler Trego methodology) will enable a clear definition of the optimum NDE capabilities and will help to identify gaps in meeting the requirements that need to be addressed with longer-term R&D. The committee believes that the validation of the selected method must include a full demonstration of the method, development of POD relationships, definition of performance limitations, and engineering parameters such as feature and component size, orientation, and accessibility.

Recommendation 30. Apply automation and data processing and data analysis technologies to augment NDE tools to perform rapid, wide-area inspections. Examples of specific technologies that should be investigated include

- effective automation of inspections and data collection equipment
- imaging technology for improved data analysis and interpretation
- data integration for different test methods for more complete and quantitative interpretation of the measurements
- scanning and automated inspection facilities, especially for ultrasonic and eddy current methods
- supportable instrumentation and equipment packaging that is convenient for the operator and can survive the depot environment
- effective and focused engineering of field equipment capable of reproducing laboratory and production test performances

Long-Term Research and Development

Recommendation 31. Develop an integrated quantitative NDE capability based on life-cycle management principles. Examples of specific tasks include

- development of probes and techniques based on accept–reject requirements dictated by the performance and materials requirements of the aircraft structure
- development and application of predictive reliability models that consider part geometry, performance requirements, NDE capabilities, failure modes, and life-cycle cost predictions
- development and application of validation and qualification techniques for NDE probes and systems using simulation models with confirmation on service components
- development of inspection standards, including reference standards using simulation techniques, to aid implementation across the entire force structure
- explore and develop the use of NDE simulation capabilities coupled with new synthetic environment (virtual reality) technology for method development and operator training

An integrated NDE program must recognize the interdisciplinary nature of NDE and the aging aircraft problem. The life-cycle approach provides a format for the development of appropriate NDE techniques that consider the performance and material requirements of the aircraft structure (accept–reject criteria), failure modes and growth characteristics (e.g., corrosion and fatigue) that contribute to the inspection interval requirements, predictive reliability models that depend on performance requirements (stress loads), NDE capabilities, and possibilities for life-cycle cost predictions. The committee believes that such an approach is key to the successful NDE management of aging aircraft. Recommended actions in other chapters of this report (characterization of corrosion rates and analytical WFD models in particular) are important to this effort. The effort must be focused and will include a mix of first-principle research, the development of probes and techniques, demonstration and validation of principles, and the development of inspection standards for implementation across the force.

The Air Force should emphasize development of new probes and techniques that address prevalent aging aircraft needs (e.g., quantitative measures of fatigue crack size and the loss of material due to corrosion). Approaches should be selected through cooperative collaboration with ALC personnel. As a guide only, crack sizes in the range of 0.030 to 0.040 in. are believed to be critical for WFD considerations, and corrosion losses in the range of 5 to 10 percent of material structures need to be detected with confidence in hidden locations (generally second- or third-layer structures) and for complex geometry. It is important that any planned probe/technique development include the specification of the POD and flaw sizing capabilities.

One approach for NDE system development is model-based computer simulation. In this approach the entire inspection process is modeled and a simulation is produced that includes part geometry, flaw characteristics, inspection modality, and analytical estimates of the system POD. These model-based simulation capabilities have been developed by interfacing descriptions of detailed part geometry from CAD data files with "measurement" models of NDE processes. Currently, measurement models exist for general ultrasonic (Coffey and Chapman, 1983; Gray and Thompson, 1986), eddy current (Nakagawa, 1988; Nakagawa and Beissner, 1990), and radiographic (Xu et al., 1994; Elshafiey and Gray, 1996; Bellon et al., 1997) applications. Measurement models enable prediction of the NDE instrument response to a flaw icon placed in various locations in the part geometry. POD can also be calculated for various flaw conditions and complex part geometry using the simulated system responses (Thompson and Schmerr, 1993). In addition, POD maps can be prepared and inspectability problems identified using these simulations.

Advances have also been made in the integration of NDE systems with structures life-cycle management (Schmerr and Thompson, 1994). For example, POD requirements for a given NDE inspection are established based on maximum allowable flaw sizes, which are determined from performance requirements, mechanical response characteristics, and material properties. In turn, inspection interval requirements are largely set by NDE inspection capabilities (particularly POD), damage growth rates (e.g., fatigue crack growth) under expected operating conditions, performance requirements, and material fracture properties (e.g., critical crack sizes). With advances in simulation capabilities, an integrated analytical approach that includes NDE measurement models, descriptions of part geometry, structural analysis codes, and damage growth predictions can be developed. The committee believes that the integrated analytical approach could be a cost-effective tool to manage the NDE inspection process for aging aircraft.

Validation techniques that use a combination of simulation techniques and limited samples to confirm validation results for NDE probes and systems should be developed and demonstrated. This task implies the development of reference standards using simulation techniques combined with a few samples to confirm validation results. The development of reference standards using simulation techniques would be a major advance.

Finally, the committee recommends that the Air Force explore and develop the use of NDE simulation capabilities using NDE measurement models and new synthetic environment (virtual reality) technology for method development and operator training. The goal of this work is to simulate complex component geometry, structures, flaws, and NDE capabilities to guide efforts to optimize equipment and sensor probe design, development of complex scan plans, and other inspection methods based on conditions expected in the depot environment.

Recommendation 32. Explore, select, and develop candidates for hybrid inspection technologies that use multiple techniques simultaneously. Examples of specific tasks include

- development of appropriate methods and models to normalize and fuse inspection data from two or more different inspection probes
- development of methods to statistically combine results and determine the POD of the hybrid system

The purpose of hybrid inspection approaches is to increase the probability of flaw detection in components with complex geometry, including hidden corrosion and fatigue cracks associated with aging aircraft. Such an approach may be required in difficult inspections that involve multiple layers. The scope of this work should include the development of appropriate theories and models to normalize and fuse inspection data from two or more inspection probes that may follow different physical measurement principles (e.g., ultrasound and eddy currents), ways to statistically combine the results and determine the POD of the hybrid system, and ways to quantify and validate the system. It is important that previous work in other areas be reviewed and adapted where possible to the aging aircraft problem. The utilization of the computer simulation models described above should be very helpful in designing and qualifying the hybrid system. Before undertaking these developments, collaborative discussions including both researchers and ALC inspectors should be pursued to define specific application areas and geometry. Some of the research required in this recommendation will be generic but some will be specific to a particular aircraft.

Recommendation 33. Perform basic and applied research to develop suitable NDE techniques to measure the integrity of composite repairs of metallic structures. Examples of specific tasks would include

- determination of the properties of the repairs (e.g., adhesive bond quality, environmental degradation, and metal substrate and repair material integrity) that need to be evaluated using NDE
- determination of appropriate accept–reject criteria and standards

The committee recommends that work be pursued at both the basic and the applied levels aimed at the development of suitable NDE techniques to measure the integrity of composite repairs on metallic structures. Although there are current limited efforts on this topic, the efforts should be increased and focused into a coordinated interdisciplinary effort. Some rather basic questions should be answered as a part of this effort and will probably require a joint effort between NDE and structures/materials researchers. One of these is the determination of the properties of the repairs that need to be tested in NDE. Examples of possibilities include the quality of joining methods (e.g., adhesive bonds, mechanical, etc.), environmental degradation at repair edges, and base metal and repair material integrity. Related and key to this question is the determination of the appropriate NDE accept–reject criteria to be applied with the NDE test to determine the state of the repair.

Recommendation 34. Develop signal and image processing techniques, applying such technologies as expert systems, neural networks, and database methods that could be used by aircraft maintenance facilities to interpret and track damage development and maintenance trends.

Work should be pursued to explore and develop useful signal and image processing techniques, applications of expert systems, using, for example, neural networks or database methods that can be used conveniently in depots and other maintenance organizations to interpret and track damage development and maintenance trends. These improvements should be targeted both to single probe inspection procedures as well as to the hybrid multimode approaches.

Recommendation 35. Increase R&D efforts for the automation of successful inspection methods and for overall automation of extensive wide-area inspections. These efforts would include two principal components:

- a generic effort based on the broad-based enhancement of scanning technology including on-board transducer (probe) mountings, signal processing methods, display techniques to enhance operator interactions, and data fusion procedures
- an effort aimed at specific aging aircraft structures and the scanning geometry needed for their inspection

The potential advantages of automated NDE include the enhancement of inspection reliability and speed through the removal of the human operator, the likelihood of reduced inspection times, and the likelihood of reduced costs. General features of scanners that need to be considered include portability, flexibility (i.e., ability to run on horizontal, vertical, and curved surfaces), ability to handle a variety of inspection modalities, and possibilities for handling hybrid multi-inspection techniques with associated signal processing and read-out procedures. The committee recommends that collaborative planning between the ALC users and researchers be in hand before and during work in this area.

Recommendation 36. Perform basic and applied research to develop suitable NDE techniques for the early detection of corrosion. Examples of specific tasks include

- (a) development of suitable NDE techniques to assess the quality and integrity of corrosion-resistant finishes

- exploration of the potential of using NDE methods to determine the initiation and level of corrosion damage

Work in NDE development that is specifically aimed at the quality of corrosion-resistant finishes and coatings has been limited. Emphasis should first be placed on understanding the ways in which finishes and coatings protect the base metal from corrosion (as recommended in Chapter 7), and with that, techniques devised to measure the degradation and failure of the protective mechanism. The Air Force Office of Scientific Research had basic materials and NDE efforts in progress, but this effort is no longer funded.

Efforts to develop NDE methods to detect the initiation of corrosion should be coupled to the development of a mechanistic understanding of corrosion and the corrosion process as presented in Chapter 7. Particular emphasis should be made to identify material parameters or properties that can be measured in service that relate to the level of corrosion. For example, the elastic constants may be sensitive to the presence of hydrogen in the material that contributes to the corrosion process. As these properties are identified, NDE sensors should be developed to provide the inspection tools. This NDE approach, if successful, would potentially provide early warning and large cost benefit to the aging fleet. This effort should be performed in collaboration with the corrosion prevention and control recommendations in this report.

MAINTENANCE AND REPAIR

Air Force research in repair technology includes R&D tasks over a broad range of topics. The primary emphasis is on the maturation of bonded composite patch repairs, especially for metallic structures. These repair methods have had successful application at the depot level (e.g., to repair fatigue cracks emanating from weep holes in C-141 lower wing skins). However, the common use of bolted repairs for both battle damage and fatigue cracking problems cannot be overlooked. In many cases bolted repairs are expected to perform well beyond their original intent, making the repair an aging structure much like the airframe itself.

The current Air Force R&D program on repairs includes

- basic research involving modeling of composite patch repairs as crack arrestors in aircraft and design and analysis techniques for composite patch repairs
- a large amount of applied research, including projects related to bonded composite patch repairs—to investigate repair procedures, analysis methods, and design considerations—along with efforts to develop repair methods and design guide for composite structures; development of advanced life-extension techniques; development of structural life enhancement, force management, and internal and external loads handbooks;

and in-service and experimental repair data. Also included are repair efforts, including projects related to bonded composite patch repairs—to develop improved materials and processing methods, investigate analysis methods, and develop repair technology handbooks—and an effort to develop repair technology for high-temperature composites

- exploratory research, including a broadly defined effort to evaluate and demonstrate repair concepts, an effort to explore the redesign of selected structural components using advanced materials and process technology, and an effort to demonstrate life-enhancement technologies for metallic structures
- a number of small short-term projects focused on optimization and demonstration of materials and processes, repair criteria, and analysis methods for bonded composite patch repairs at the depot level; also included are projects to evaluate methods to generate stress spectrum and to evaluate cold-expansion bushing repairs

The committee believes that the focus on optimization of materials and processes and analysis tools for bonded composite repair of metallic structures is appropriate because the Air Force has unique expertise in this technology. The committee also supports the planned research focused on the redesign of components to take advantage of advances in materials and processing technology.

Although the current R&D program in the area of repairs is well planned, there are no current programs in the repair task that consider the removal, surface treatment, and reapplication of corrosion-resistant finishes or protection systems. This is a particular shortfall considering the materials and process changes that will be necessitated by environmental regulations concerned with the elimination of heavy metals (e.g., chromium and cadmium) and limits on volatile organic releases. The Air Force is currently undertaking a great deal of research on environmentally compliant finish material and process development (Donley, 1996), but has not yet come to terms with the particular needs of aging aircraft in this area.

In general, the committee believes that the concept of repairs should be expanded to include maintenance and repair. This change would require closer coordination of R&D tasks in repair with NDE tasks and an emphasis on implementation of developed technology through the development of generic repair design and processing handbooks and engineering analysis tools to broaden the application of new repair technologies.

The committee recommends that the emphasis of the repair R&D programs be increased in the following areas:

- technologies for the removal, surface preparation, and reapplication of corrosion-resistant finishes
- evaluation guidelines for the relative lives of bolted repairs

- guidelines for taking advantage of advances in materials and processing technology in component replacement (including an examination of certification requirements to see if they can be waived or simplified without compromising safety); an example would be to reduce susceptibility to stress corrosion cracking through the use of improved aluminum alloys, tempers, and processes to reduce residual stresses
- repair and analysis methods for maintenance of structures susceptible to high-cycle fatigue
- maintenance and repair methods and guidelines for advanced composite structures

Near-Term Research and Development

Much has been learned in the past ten years concerning methods to analyze and repair damage in primary metallic and composite structures. Although the focus of much of the early work was on designing repairs for battle damage, the focus more recently has been on repairs for durability and life extension for current aircraft. The primary focus of the near-term programs for aging aircraft must be to identify the lessons learned from recent programs (e.g., C-141 and battle damage repair) and apply them at the maintenance organizations where they can be used to make the repairs that can extend the life of current aircraft.

Recommendation 37. Develop tools and guidelines to implement recent advances in bonded repair of primary structure for Air Force and contractor maintenance organizations. Examples of specific tasks include

- optimization and validation of materials and processes, including adhesive materials and surface preparation and bonding processes
- development of computational tools and guidelines for the design and analysis of design bonded repairs
- validation and documentation of analysis techniques to evaluate continuing damage growth beneath bonded repairs (CALCUREP) and bolted repairs (RAPID)

To ensure that structural repairs have the best possible chance for success, the committee recommends that materials and processes that have been developed to join the repair to the structure, seal the repaired surface from further degradation due to adverse environments, and protect the repair from rapid deterioration in the flight environments be documented and made available to the maintenance organizations. Materials and processes to be considered include surface preparations, adhesives, and bagging materials used for successful repairs of the C-141. Advances in these material systems and any new, validated processes must be demonstrated by maintenance personnel with on-site consultation from the developing organization.

There are a number of design and analysis tools for repairs that have been developed in the recent past (Bakuckas et al., 1996; Fredell et al., 1996). Once validated these methods will provide the ALCs far better and faster means to design reliable repairs than those currently in use. Design and analysis tools must include capabilities in the following areas to be used in the design of reliable and durable repairs: (1) continuing damage growth beneath the repair due to fatigue loads, (2) reliability and durability of bond or bolted joints, and (3) variations in repair materials and processes used to fabricate and apply the repair. Although analysis codes such as A4EI, PGLUE, and RAPID perform analyses of bonded or bolted repairs, they are very limited in the types of repair geometries to which they are applicable. A4EI applies only to a linear bonded repair, PGLUE to doubly symmetric bonded repairs, and RAPID to bolted repairs. There is much to be done to extend these methods to explicitly analyze realistic three-dimensional structures.

The growth of damage beneath the repair is a critical concern. Bonded composite repairs are intended to provide sufficient stiffness and constraint of the structure so that the stress intensity factors for existing flaws are reduced to levels below threshold so that they cannot continue to grow. Analysis routines such as those in the current version of CALCUREP (for bonded repairs) and RAPID (for bolted repairs) need to be validated to ensure their accuracy and then be made available to the ALCs.

Recommendation 38. Develop analytical tools to take advantage of effective solid model interfaces developed for finite element modeling to model and simulate repair methods and geometric relationships for particular component repairs.

Methodology has been developed, under Navy funding, that uses super-element technology to allow limited use of vehicle-level finite element model analyses on laptop PC hardware (Goering and Dominguez, 1992). With condensation techniques to reduce the degrees of freedom within the model, it is possible to design sophisticated large-scale repairs of damage to major structural members, to assess structural integrity before and after repair, and to assess the feasibility and capability of the repair to restore the structure to its original function.

With the visualization possible on laptop PCs to provide a three-dimensional image of the area to be repaired, the loading conditions, and the damage to be repaired, the current capability to perform rapid repair analysis is remarkable. Unfortunately, the modeling of such repairs is still a time-consuming process. Work needs to be performed to make automatically generated repairs for a number of typical damage scenarios available. Although this initial effort might be limited in what it can provide, it could be a valuable tool for maintenance organizations.

Recommendation 39. Develop and validate guidelines for the relative lives of bolted repairs. Specific tasks include

- testing to evaluate crack stopping by cold working, peening, laser shock treatment, stop drilling, or hole filling
- testing to evaluate repair designs, including issues such as protection systems, taper ratios, fastener patterns, and fastener types
- testing to evaluate innovative fastener concepts such as single-shank blind fasteners and hole-expanding blind fasteners

Bolted repairs are the most common repair applied to aircraft structures. Their capability to extend lives is limited because bolted joints tend to loosen up and the load transfer occurs away from the damaged area. Like bonded repairs, bolted repairs provide the reduction in strain levels at the damage site. However, neither repair system is expected to provide restoration of strength in damaged structure to the original design loads for the life of the airframe. Bolted repairs are generally expected to extend lives of damaged structures to the next programmed depot maintenance cycle. However, experience indicates that the repairs are often called upon to remain effective in providing structural integrity far longer than a single depot maintenance cycle. In such cases, determination of the relative lifetimes for several bolted repair configurations is desirable so that any selection of repair configuration will take into consideration the lifetime requirement and repair capability.

Bolted repairs are limited by the limited fatigue life of the blind fasteners typically used to install these repairs from one side of the closed box structures. The development of blind fasteners with improved fatigue lives, either through improved design or through interference in the hole, would provide significant benefits to the life of the repair attachment.

In addition, there are a number of methods to extend the lives of the damaged structure: through cold working, peening, laser shock treatment, or hole filling. The ability of these treatments to provide extended lives must be verified and quantified by test.

The techniques described above should be incorporated into design methods for repairs that assure, through damage tolerance analyses and verified by test, that the repair will retard or stop the flaw growth from previous damage. Moreover, the design must be sensitive to the potential for the development of flaws in the structure surrounding the repair since the load distributions nearby have been changed by the repair. Taper ratios and fastener pattern designs, along with fastener sizing for flexibility and strength, can provide significant life improvements for bolted patches, but test data must verify the projected improvements.

Recommendation 40. Develop guidelines and procedures to reduce the cost of accepting new materials and structures as replacements for aging structural components.

Since the design and manufacture of many of the aircraft that constitute the aging force, significant advances have been made in materials and processing technology to improve the resistance of aircraft components to aging degradation. For example, corrosion and stress corrosion cracking (SCC) resistance can be significantly upgraded through the use of substitute materials and heat treatments (e.g., more-corrosion-resistant 7050, 7150, or 7055 alloy for 7075, SCC- and exfoliation-resistant T-7X tempers for 7XXX-series aluminum alloys), improved protective finishes and corrosion prevention compounds, and incorporation of design features such as drainage and sealing to prevent corrosion. However, advances in materials and process technology have not been captured because of the excessive cost and time required to qualify them for service and because of the long lead times required for small-quantity procurement. Currently, material substitutions are handled on a individual part-by-part basis. The committee recommends that the Air Force develop guidelines to facilitate the force-wide implementation of the best materials and processing solutions while minimizing evaluation and qualification test requirements. Examples of specific tasks include

- substantiation of improved materials as preferred replacements for SCC- and corrosion-susceptible alloy components
- development of an approved alloy substitution matrix
- evaluation of test protocols for replacement materials and structures to allow for one-time approval of general materials substitutions

This effort would reduce test costs for replacement structures, but would also act as an incentive to replace older, more-damage-prone materials with more-damage-resistant materials. Considered separately, the quantity of material required for validation efforts and support of replacement modification programs is small. However, quantities required for more general materials substitutions could be significant enough to enable reduction of long lead times associated with small-quantity procurement by stocking qualified substitutes.

Recommendation 41. Develop repair design guidelines for dynamically loaded structures. Examples of specific tasks would include

- documentation of repair materials and processes and design considerations based on an understanding of root causes, dynamic load conditions, and environmental factors

- develop and validate damped repair concepts based on currently available adhesive and composite repair technology

Repairs for dynamically loaded structures offer the unique potential to significantly reduce load magnitudes or change the critical load frequencies while they serve to recover the integrity of the structure. The challenge for repairs of dynamically loaded structures is to recover the structural integrity and stiffness requirements while not moving critical dynamic modes into surrounding structures where damage can occur even more rapidly than in the initial failure. This is why knowledge of the dynamic modes and responses of both the original structure and the repaired structure are so important to the repair of dynamically loaded structures.

Recently, adhesives that contain significant damping potential have become available with sufficient durability that they can be used in bonded repairs. These adhesives, combined with stand-off materials to maximize the shear transfer through the adhesive and composite skin materials to withstand low-velocity impacts and provide load-carrying capability, have provided the opportunity to design and fabricate repairs that damp the loads that cause high-cycle fatigue failures. Before these repairs can be used with confidence by the Air Force maintenance organizations, they must be verified to provide continuity of the structure while reducing the driving forces below that level which initiates failures in a part for the remainder of its design life.

Long-Term Research and Development

Recommendation 42. Develop an expert system to aid in the assessment of damage, the need for repair, and the design and analysis of repairs.

The committee believes that an expert system should be developed that has the capability to recall vehicle level loads and structural analysis, graphically isolate the region being repaired, and assess the viability, reliability, and durability of the repair. These systems would use databases developed and maintained by the recommended corrosion and fatigue working groups discussed in Chapter 5. Analysis methods should be developed that are capable of analyzing bolted or bonded joints for real repair configurations in which existing fastener patterns and other structural details need to be accommodated. Some of these more-flexible analysis tools have been developed, but are cumbersome and time-consuming to use. Simplifications in graphical interfaces and the ability to handle large data files representing complex three-dimensional geometries may permit better interfaces between structure and repair to be developed. It is possible to envision a virtual repair routine for a laptop environment that could lead the repair technician or analyst through the steps of the repair by providing both graphical and descriptive specifications of the repair processes and procedures.

Recommendation 43. Develop a common database of repair lessons learned, to be managed and maintained by the maintenance and repair working group, that would be available to the ALCs and would contain information on repair experience, including both adverse and successful results, problems in assessment, design, analysis, materials, or application of the repair.

Recommendation 44. Develop analysis methods and life prediction tools and methods for structural repairs and affected structure.

There are several methods for the analysis of bonded patch repairs. They can be classified broadly as either analytical or numerical. The analytical approach of Rose (1981) is based on Hart-Smith's (1974) theory of bonds, elastic inclusion analogy, and on some simplifying assumptions. Fredell (1994) has extended this analysis to include thermal effects and has also carried out an evaluation of mechanical doubler repairs. Erdogan and Arin (1972) have used an integral equations approach to study bonded patch repairs. The assumptions of Erdogan and Arin were subsequently used by Ko (1978) and Hong and Jeng (1985) in an analysis of sandwich plates with part-through cracks.

Jones and Callinan (1979), Mitchell et al. (1975), and Chu and Ko (1989) have used the finite element method to study bonded patch repairs. Park et al. (1992) have used an integral equation approach combined with the finite element alternating method to estimate the stress intensity factors for patched panels. Tarn and Shek (1991) have combined the boundary element method (for the plate) and finite element method (for the patch) to estimate the stress intensity factors. Other work in this area includes Atluri and Kathiresan (1978), Sethuraman and Mathi (1989), and Kan and Ratwani (1981). A comprehensive summary of the analytical and numerical work on composite patch repairs appears in a recent monograph (Atluri, 1997).

In most of these approaches, only patches of infinite size, very narrow strip-type patches, or infinite sheet cases are considered. All of these cases are valid only for flat sheets. The loading for these analyses are hoop stresses evaluated from basic thin-shell theory. Although in most cases this is a good approximation, this does not take into account the stress redistribution due to curvature and to the presence of stiffeners.

Specific capability improvements that are needed include the ability to analyze the following structural details:

- the joint between the repair and the original structure
- the damaged structure with the repair in place
- the surrounding structure affected by changes in load paths
- complex and curved structural details

Recommendation 45. Develop, characterize, and evaluate improved damping materials with improved environmental resistance and low-temperature performance for repair and modification of dynamically loaded structures. Examples of specific tasks include

- development of accelerated environmental test methods and criteria to evaluate resistance to aircraft service conditions, including thermal and fluid exposures
- development and validation of repair concepts that include improved damping materials

Damping material systems currently in use have shown inadequate durability. The committee recommends that long-term research be conducted to develop improved damping material systems that provide low-temperature damping performance and better resistance to aircraft fluids and environmental exposure. Candidates should be tested under low-temperature conditions, with temperature cycling through realistic aircraft environments, including moisture and fuel, where necessary. Methods to accelerate this type of testing will be important for both the screening of developmental systems and for the characterization and acceptance of selected systems.

Repair designs that use these improved damping systems must be validated to ensure that the improved performance translates into more-durable repairs. These systems may require additional care to ensure their durability. Damped composite repairs provide the potential to seal the stand-off material to prevent or delay moisture intrusion. Best practices must be incorporated into the repair system to ensure the integrity of the bond and the effectiveness of the damping materials.

9

Prioritized Research Recommendations

Because of the budget pressures and difficult choices associated with conducting and managing a R&D program, the committee task included a charge to prioritize research recommendations. The committee developed criteria that were used to prioritize all of the research recommendations in Chapters 6 to 8.

Priority levels for recommended R&D opportunities were established relative to the Air Force objectives introduced in Chapter 1 (i.e., safety of flight [Objective A], maintenance costs and force readiness [Objective B], and economic life estimates [Objective C]). Clearly, research that eliminates substantial threats to flight safety receives consideration for the highest priority to the Air Force. However, the escalation of maintenance costs and the impact on force readiness has become a pervasive issue that, if allowed to continue unchecked, could significantly hamper the ability of the Air Force to field a force that meets mission requirements for capability and readiness. The committee did not prioritize the recommendations with respect to Objective C because they found that research recommendations to develop technology to support economic life estimates related closely to the more important Objective B.

Definitions of priority categories for near-term (to support near-term engineering actions in the next five years) and long-term (more than five years until implementation) R&D recommendations include

Critical priority: essential to flight safety (Objective A) (i.e., would eliminate a substantial threat to flight safety)
Priority 1: essential to the reduction of maintenance costs and improvement of force readiness (Objective B) (i.e., would enable the Air Force to address significant technical problems)
Priority 2: important to improved flight safety (Objective A) or reduced maintenance costs and improved force readiness (Objective B) (i.e., would represent significant improvements over current solutions)
Priority 3: advantageous to improved flight safety (Objective A) or reduced maintenance costs and improved force readiness (Objective B) (i.e., would improve efficiency or reduce cost of current methods)

In addition, the committee assigned technical risk categories for long-term research recommendations. Technical risk is an assessment of the difficulty involved in achieving a technical objective. The committee designated technical risk associated with long-term research opportunities as low (validation and implementation of technology that is essentially mature), moderate (some further technology development and scaling required), and high (significant technology advancement required). The long-term research program should contain a mix of risk categories. Moderate- and high-risk programs should be monitored closely by the proposed aging aircraft technical steering group to ensure continued progress in clearing technical hurdles and continued need for the resulting technology for the maintenance of the aging force. Near-term opportunities were generally assumed to have low technical risk.

CRITICAL PRIORITIES

There are no research efforts identified at this time that are considered of sufficient magnitude to be categorized as critical priorities. However, the committee believes that it is possible that the durability and damage tolerance updates recommended in Chapter 5, and in particular the high-priority updates on the F-16, A-10, U-2, and T-38 aircraft, will identify critical priority near-term research and engineering tasks. These could include

- development of specific inspection instruments or procedures
- development of specific repair or modification designs or processes
- development and use of more sophisticated analysis procedures and additional full-scale fatigue testing to identify fatigue-critical areas
- obtaining additional flight loads and environment data for specific aircraft

NEAR-TERM RESEARCH

Prioritized recommendations for near-term R&D are shown in Table 9-1, including the recommendation number, a brief description of the recommendation, the page where the recommendation is discussed, the objective that is addressed primarily by the recommended research, and the suggested

TABLE 9-1 Prioritized Near-Term Research Recommendations

No.	Recommendation	Description	Objective	Priority
	Fatigue			
(1)	Fail-safe residual strength prediction methods	Page 50	A	2
(2)	Improve current methods to estimate the onset of WFD	Page 50	A	2
(6)	Methods to predict dynamic responses	Page 52	B	2
(11)	Effect of corrosion damage on material properties	Page 55	A	3
(12)	Effect of corrosion and corrosive environment on safety limits	Page 55	A	3
(13)	Effect of joint pillowing on fail-safety	Page 55	A	2
	Corrosion Prevention and Control			
(17)	Laboratory test protocol for accelerated corrosion testing	Page 57	B	2
(18)	Evaluate durability of new protective coatings	Page 58	B	1
(19)	Methods for early detection of corrosion	Page 58	B	2
	Stress Corrosion Cracking			
(23)	Affordable upgrades in SCC prevention	Page 60	B	1
(24)	Evaluation of SCC protection systems	Page 60	B	1
(25)	Residual stresses and their alleviation	Page 61	A	2
(26)	SCC susceptibility of Air Force alloys	Page 61	A	2
	NDE			
(29)	Evaluate, validate, and implement NDE equipment and methods	Page 64	B	1
(30)	NDE automation, data processing, and analysis	Page 66	B	2
	Maintenance and Repair			
(37)	Guidelines to implement advances in bonded repairs	Page 69	B	2
(38)	Solid model interfaces to simulate repair methods	Page 70	B	2
(39)	Guidelines on relative lives of bolted repairs	Page 70	A	3
(40)	Reduce cost of materials and structures substitution	Page 71	B	2
(41)	Repair design guidelines for high-cycle fatigue problems	Page 71	B	2

priority. Priority 1 recommendations include (1) research to develop and implement corrosion prevention and control procedures and (2) evaluation and implementation of nondestructive evaluation techniques that address specific Air Force aging aircraft issues.

LONG-TERM RESEARCH

Prioritized recommendations for long-term R&D are shown in Table 9-2, including the recommendation number, a brief description of the recommendation, the page where the full recommendation appears, the objective that is addressed primarily by the recommended research, an assessment of technical risk, and the suggested priority. Priority 1 recommendations include (1) research to develop a fundamental understanding of corrosion and stress corrosion cracking to support the development of improved corrosion prevention and control procedures and (2) development and validation of rapid, wide-area nondestructive evaluation techniques to address specific aging aircraft needs.

PRIORITIZED RESEARCH RECOMMENDATIONS

TABLE 9-2 Prioritized Long-Term Research Recommendations

No.	Recommendation	Description	Objective	Technical Risk	Priority
	Fatigue				
(3)	WFD crack formation and distribution relationships	Page 50	A	moderate	2
(4)	Analytical prediction of WFD crack distribution functions	Page 51	A	high	2
(5)	Validation of analytical WFD methods	Page 51	A	low	2
(7)	Crack growth threshold behavior	Page 52	B	low	2
(8)	Analytical methods to predict dynamic behavior	Page 53	B	moderate	2
(9)	Expert system for high-cycle fatigue repairs	Page 53	B	high	3
(10)	Dynamic load monitoring and alleviation	Page 53	B	moderate–high	2
(14)	Effect of environment on growth of small cracks	Page 55	A	low	2
(15)	Effect of flaw morphology on crack growth	Page 56	A	moderate–high	2
(16)	Effect of hydrogen on fatigue crack growth	Page 56	A	moderate	3
	Corrosion Prevention and Control				
(20)	Basic research in corrosion prevention and control	Page 59	B	high	1
(21)	Corrosion rates for major corrosion types	Page 59	B	moderate	2
(22)	Basic research in coating durability	Page 60	B	moderate	1
	Stress Corrosion Cracking				
(27)	Fundamental research in SCC prevention	Page 61	B	moderate–high	1
(28)	Life prediction methods for SCC	Page 62	B	high	2
	NDE				
(31)	Develop integrated quantitative NDE capability	Page 66	B	moderate–high	1
(32)	Hybrid inspection technologies	Page 67	B	high	2
(33)	NDE to assess composite repairs	Page 67	B	high	2
(34)	Advanced technologies to track maintenance trends	Page 68	B	moderate–high	3
(35)	Automation of wide-area inspections	Page 68	B	moderate	1
(36)	NDE for early corrosion detection	Page 68	B	high	3
	Maintenance and Repair				
(42)	Expert system for design and analysis of repairs	Page 71	B	moderate	2
(43)	Common database of repair lessons learned	Page 72	B	low	2
(44)	Analysis methods for structural repairs	Page 72	B	moderate	3
(45)	Damping material for dynamically loaded structures	Page 72	B	moderate	3

10

Future Structural Issues: Composite Primary Structures

The issues and concerns identified by the committee during this study have concerned metallic alloy structures that make up the vast majority of the airframes in the Air Force aging aircraft. However, more-recent aircraft have significant quantities of the flight control (C-17) and primary airframe structures (B-2, F-22) constructed from carbon-fiber-reinforced polymeric composites. The purpose of this chapter is to describe service experience with composite structure—including Air Force and commercial experience with secondary structures and flight control structures as well as Navy and commercial experience with primary structure—and to recommend potential degradation mechanisms and actions to monitor and evaluate the aging of composite structures in the future.

APPLICATIONS AND SERVICE EXPERIENCE

Secondary Structures

The application of polymeric composites has been an evolutionary process, with increased use as materials and processing technology matured and program needs dictated their use. First-generation, glass-reinforced composites, in the form of thin-facesheet honeycomb sandwich constructions, have been in general use for secondary structures (i.e., wing-to-body fairings, fixed-wing and empennage cover panels, and secondary control surfaces) on Air Force and commercial transport aircraft since the 1960s.

During the 1970s, the commercial availability of carbon and aramid fibers, the performance enhancements made possible by weight savings, and uncertainty in fuel supply and costs provided an impetus for the development and application of structural composites for airframe applications. The Air Force conducted much of the pioneering research in materials, processes, and design of composite structures leading to the application of composites in secondary and flight control structures on the F-15, F-16, and B-1B. The materials used for these components included largely unmodified amine-cured epoxy resins (e.g., TGMDA/DDS) reinforced with aramid (Kevlar® 49), carbon (e.g., Amoco T-300, Hercules AS-4), and E-glass fibers. Structures were generally thin 0.6- to 1.5-mm (0.024- to 0.060-in.) facesheets co-cured or secondarily bonded to composite or aluminum honeycomb core.

At about the same time, the commercial industry became interested in the application of composite structures. To encourage the use of composites in commercial production applications, NASA conducted technology development and flight service programs, including design, certification, and use in airline service. Carbon/epoxy, aramid/epoxy, and aramid-carbon/epoxy and glass-carbon/epoxy hybrid composites were first used on a production scale in the early 1980s for the generation of aircraft that included the Boeing 757, 767, and 737-300; Airbus A310 and A320; and McDonnell Douglas MD-80 series. Applications included secondary structures such as fairings, fixed-wing and empennage cover panels, and engine cowlings, as well as primary flight controls such as ailerons, elevators, rudders, and spoilers. The number of aircraft involved and the high use rates of commercial aircraft has led to a great deal of service experience with composites in commercial aircraft applications. For example, NASA has conducted flight service evaluations of 350 components with over 5.3 million total flight hours (Dexter and Baker, 1994).

In general, the service experience with composites indicates that damage occurs because of discrete sources such as impacts, lightning strikes, and handling rather than progressive growth caused by fatigue conditions (NRC, 1996a). The types of damage to composite components include disbonds or delaminations, holes or punctures, cracks, and other damage. An especially difficult maintenance issue resulting from these types of damage is when perforation of the facesheet allows hydraulic fluids, water, and other liquids to move into the honeycomb core.

Primary Structures

Throughout the 1970s and 1980s, the Air Force was instrumental in the development of materials, processes, and design considerations for primary structural applications of polymeric composites. The Air Force has only recently made significant use of composite primary structure on the B-2 and will continue on the upcoming F-22. The Navy and the commercial aircraft industry have limited service experience for primary composite structures on the Navy F/A-18 and AV-8B and on the Airbus A320. The constructions are integrally stiffened carbon/epoxy laminate skin fabricated from materials similar to the first-generation materials previously used for secondary structure and primary flight controls. The further development of carbon fibers with improved strength and modulus (e.g., Hercules IM7 and Toray T-800H) and high-performance and toughened matrix polymers has led to application on the Boeing 777 empennage to expand the primary structural applications.

Guidance for the selection, design, and analysis of composite structures for polymeric composites have been developed over the past 25 years (Whitehead et al., 1986; Vosteen and Hadcock, 1994). These methods, forming the basis for MIL-HNBK-17 (DOD, 1994), are based on static ultimate strength considerations and the effects of three primary structural degradation mechanisms:

- *Impact damage.* To verify impact tolerance, the structure is subjected to a low-velocity impact prior to the fatigue testing to substantiate inspection intervals and performance for the life of the structure under barely visible impact damage criteria.
- *Mechanical fatigue.* Fatigue is not generally a significant damage mechanism in fiber-dominated composite structures that meet impact damage tolerance requirements described above (Jeans et al., 1980). Components that experience significant interlaminar or out-of-plane loading can be susceptible to fatigue damage.
- *Humidity (or fluid) exposure.* Design properties based on coupon tests are typically generated in a fully saturated humidity condition (85 percent relative humidity). Real-time exposures, using flight service components and ground exposures, have verified this approach (Dexter and Baker, 1994).

Consideration of these degradation mechanisms and the use of structural design verification testing, with evaluations on scales from coupon level to full scale, have successfully offset the limitations of design analysis methods in terms of prediction of interlaminar stresses, damage initiation, and delamination growth (NRC, 1996b). The final step of this approach is typically a full-scale component fatigue test on an impact-damaged structure.

The limited experience of the Navy and commercial aircraft service with composite laminate constructions used for primary structures has been good. There have been very few occurrences of damage in primary structures, and in the few cases that have occurred, there have been no indications of progressive damage. Potential degradation mechanisms to monitor in the future for composite structural applications include (1) the development of transverse matrix cracking due to mechanical, thermal, or hygrothermal stresses; (2) the growth of impact damage under fatigue loading; (3) the growth of manufacturing-induced damage, especially from fastener installation; and (4) the development of corrosion in adjacent metal structures.

RECOMMENDATIONS FOR LONG-TERM RESEARCH

The committee recommends that the Air Force undertake research to monitor potential deterioration of composite structures and to develop or improve maintenance and repair technologies, especially for composite primary structures. The recommendations are considered long-term research opportunities because they do not specifically support near-term engineering or management actions discussed in Chapter 5.

Recommendation 10-1. Develop, validate, and implement NDE equipment and methods to reliably detect degradative damage of composite structures, especially the development of transverse matrix cracks, impact damage, delamination associated with growth of manufacturing-induced damage around fasteners, moisture penetration near edges, and corrosion of adjacent structure.

The committee recommends that the Air Force evaluate, adapt, and utilize NDE advances to develop methods and equipment capable of evaluating the key composite damage mechanisms. Emphasis should be placed on automated methods, compatible with depot-level application, to perform rapid, wide-area inspections. As described in Chapter 8, the committee recommends a life-cycle approach to evaluate and validate methods that considers detectability and inspectability, full-scale validation, material degradation mechanisms, technique reliability, inspection intervals, and cost. The most promising technologies that are currently available include ultrasonic methods (c-scan, scan imaging, and resonance techniques) and thermal methods (large-area impulse heat technique). There have been significant advances in automated inspection methods for production and in-process inspection of composite structures that could be adapted to the depot environment.

Recommendation 10-2. Develop tools and guidelines to standardize bonded repair methods for composite structures.

Occasionally, temporary or permanent repairs of composite honeycomb structures can be performed by bonding or bolting a sealant-coated metal or precured composite overlay over the damage. However, most permanent repairs are accomplished with room-temperature curing wet lay-up, precured patch, and elevated temperature prepreg repair techniques. The Air Force has a unique capability, as described in Chapter 8, in the area of laminated composite patch repairs for metal structures. The techniques and tools developed for the design and evaluation of repair of aged metallic structures should be extended and validated for composite structures.

Perhaps the most pressing problem in patch repairs of composite structures is that the structures are fabricated from a large number of resin/reinforcement systems from several qualified suppliers, requiring the repair depot to stock a variety of repair materials. There is a pressing need to standardize repair materials and processes across the Air Force inventory. The Commercial Aircraft Composite Repair Committee (CACRC) has been formed to address composite service and repair concerns of the commercial aircraft industry. The Air Force should monitor the activities of the CACRC and evaluate the applicability of the recommendations of the CACRC to Air Force aircraft.

Recommendation 10-3. Develop tools and methods for bolted repairs of composite primary structures.

The thicker laminate construction used in composite primary structures, as well as the size and nature of discrete damage from typical aircraft service (e.g., impact damage, lightning attachment damage, delaminations), are not conducive to wet lay-up patch repair technologies. Thin facesheets on honeycomb panels are generally repaired using bonded scarf patches with a scarf taper of 20:1, which, if applied to thicker laminate constructions, would result in the removal of a large amount of undamaged material (Bodine et al., 1994). Much of the emphasis in the development of primary structure repairs has therefore been on fastened, precured composite or metallic splice plates, similar to current bolted repair techniques for metal structure. The issues that must be addressed in these types of repairs include (1) criteria for determining when repairs are required; (2) availability of standardized repair elements; (3) drilled hole quality; (4) ability to restore original strength, durability, and damage tolerance; and (4) ability to match existing contours.

Recommendation 10-4. Evaluate environmentally benign paint removal methods recommended in Chapter 7 for compatibility with polymeric composite substrates.

Composites must be protected by finishes with resistance to fluid penetration and UV degradation. Maintenance of protective finishes represents significant operational costs to the Air Force. The removal of finishes from composites is a slow and expensive process. Because chemical strippers attack the polymer matrix, finishes generally are removed using mechanical abrasion processes. New paint removal processes such as laser, heat, frozen carbon dioxide blasting, and wheat starch blasting are being evaluated. Rapid, low-cost, on-aircraft paint removal techniques are needed to reduce the cost of maintaining composite structures and to preclude damage to the structure.

References

AGARD (Advisory Group for Aerospace Research and Development). 1990. Short-Crack Growth Behaviour in Various Aircraft Materials. P. R. Edwards and J. C. Newman, Jr., compilers, AGARD-R-767. Neuilly-Sur Seine, France: North Atlantic Treaty Organization.

AGARD. 1992. Environmentally Safe and Effective Processes for Paint Removal, AGARD-CP-791. Neuilly-Sur Seine, France: North Atlantic Treaty Organization.

AGARD. 1996. Environmentally Compliant Surface Treatments of Materials for Aerospace Applications, AGARD-CP-816. Neuilly-Sur Seine, France: North Atlantic Treaty Organization.

Agarwala, V.S., and A. Fabiszewski. 1994. Thin Film Microsensors for Integrity of Coatings, Composites, and Hidden Structures. Presentation at the 1994 NACE International Corrosion Conference, Baltimore, Maryland, March.

Agarwala, V.S., P.K. Bhagat, and G.L. Hardy. 1995. Corrosion detection and monitoring of aircraft structures: An overview. Pp. 19-1–19-6 in Corrosion Detection and Management of Advanced Airframe Materials, AGARD-CP-565. Advisory Group for Aerospace Research and Development. Neuilly-Sur Seine, France: North Atlantic Treaty Organization.

ASM. 1987. Metals Handbook, Vol. 13. Corrosion. Metals Park, Ohio: ASM International.

Atluri, S.N. 1997. Structural Integrity and Durability. Forsyth, Georgia: Tech Science Press.

Atluri, S.N., and K. Kathiresan. 1978. Stress analysis of typical flaws in aerospace structural components using 3D hybrid displacement finite element method. Pp. 340–351 in Proceedings of the 19th AIAA/ASME Structures, Structural Dynamics, and Materials Conference. New York: American Institute of Aeronautics and Astronautics.

Bakuckas, J.G., C.C. Chen, P.W. Tan, and C.A. Bigelow. 1996. Engineering Approach To Damage Tolerance Analysis of Fuselage Skin Repairs. DOT/FAA/CT-95/75. Atlantic City, N.J.: Federal Aviation Administration Technical Center.

Barnard, D.J., and D.K. Hsu. 1997. Development and testing of the dripless bubbler ultrasonic scanner. Pp. 2069–2076 in Review of Progress in Quantitative Nondestructive Evaluation, Vol. 16. D.O. Thompson and D.E. Chimenti, eds. New York: Plenum Press.

Beier, T.H. 1997a. Sonic fatigue—A problem requiring special handling. Pp. 597–612 in Proceedings of the 1996 USAF Aircraft Structural Integrity Conference. WL-TR-97-4055. Wright-Patterson AFB, Ohio: Wright Aeronautical Laboratories.

Beier, T.H. 1997b. Effective Buffet Load Alleviation Via an Active Control Surface. Presentation at the DoD/FAA/NASA Conference on Aging Aircraft, Ogden, Utah, July 8–10.

Bellon, C., G. Tillack, C. Nockermann, and L. Stenzel. 1997. Computer simulation of x-ray NDE process coupled with CAD interface. Pp. 325–330 in Review of Progress in Quantitative Nondestructive Evaluation, Vol. 16. D.O. Thompson and D.E. Chimenti, eds. New York: Plenum Press.

Berens, A.P. 1989. NDE reliability data analysis. Pp. 689–701 in Metals Handbook, Volume 17: Nondestructive Evaluation and Quality Control. Metals Park, Ohio: ASM International.

Bieber, J.A., S.K. Shaligran, J.N. Rose, and J.C. Moulder. 1997. Time-gating of pulsed eddy current signals for defect characterization and discrimination in aircraft lap-joints. Pp.1915–1922 in Review of Progress in Quantitative Nondestructive Evaluation, Vol. 16. D.O. Thompson and D.E. Chimenti, eds. New York: Plenum Press.

Blohowiak, K.Y., J.H. Osborne, K.A. Krienke, and D.F. Sekits. 1997. Durable Sol-Gel Surface Preparations for Repair and Remanufacture of Aircraft Structures. Presentation at the DOD/FAA/NASA Conference on Aging Aircraft, Ogden, Utah, July 8–10.

Bodine, J.B., B.W. Flynn, M.H. Gessel, L.B. Ilcewicz, G.D. Swanson, M.F. Nahan, and J.R. Epperson. 1994. Repair of Composite Fuselage Panels. Paper presented at the Fifth NASA/DOD Advanced Composite Technology Conference, Seattle, Washington, August 22–25.

Boeing. 1994. Aging Airplane Corrosion Prevention and Control Program: Model 737-100/200. Document D6-38528, revision D. Seattle, Washington: Boeing Commercial Airplane Group.

Bucci, R.B., and C.J. Warren. 1997. Material Sunstitutions for Aging Aircraft. Presentation at the DoD/FAA/NASA Conference on Aging Aircraft, Ogden, Utah, July 8–10.

Bucci, R.B., R.L. Brazill, and J.R. Brockenbrough. 1986. Assessing growth of small flaws from residual strength data. Pp. 541–556 in Small Fatigue Cracks. Warrendale, Pennsylvania: The Minerals, Metals, and Materials Society.

Bucci, R., R. Bush, A. Hinkle, H. Konish, M. Kulak, and R. Wygonik. 1996. High-strength aluminum forged product durability and damage tolerance for weight and total life cost saving. Pp. 681–702 in Proceedings of the 1995 USAF Aircraft Structural Integrity Program Conference, Vol. 2. WL-TR-96-4094. Wright-Patterson AFB, Ohio: Wright Aeronautical Laboratories.

Buchheit, R.G. in press. Copper removal during formation of corrosion resistant alkaline oxide coatings on Al-Cu-Mg alloys. Journal of Applied Electrochemistry.

Cannava, V. 1997. LogisTech, Inc. Personal communication, April 10, 1997.

Chu, R.C., and T.C. Ko. 1989. Isoparametric shear spring element applied to crack patching and instability. Theoretical and Applied Fracture Mechanics 11:93–102.

Coffey, J.M., and R.K. Chapman. 1983. Applications of elastic scattering theory for smooth flat cracks to the quantitative prediction of ultrasonic defect detection and sizing. Nuclear Energy 22(5):319–333.

Cowie, W.D. 1989. Fracture control philosophy. Pp. 666–673 in Metals Handbook, Volume 17: Nondestructive Evaluation and Quality Control. Metals Park, Ohio: ASM International.

Dexter, H.B., and D.J. Baker. 1994. Flight service environmental effects on composite materials and structures. Advanced Performance Materials 1(1):51–85.

DOD (U.S. Department of Defense). 1987. Airplane Damage Tolerance Requirements. MIL-A-83444. Washington, D.C.: U.S. Department of Defense.

DOD. 1988. Aircraft Structural Integrity Program Airplane Requirements. MIL-STD-1530, revision A. Washington, D.C.: U.S. Department of Defense.

DOD. 1994. Polymer Matrix Composites, Vol. 1. MIL-HNBK-17. Washington, D.C.: U.S. Department of Defense.

Donley, M.S. 1996. Advanced Aircraft Coatings: Program Overview. Presentation to the Committee on Aging of U.S. Air Force Aircraft, National Materials Advisory Board, National Research Council, Wright-Patterson AFB, Ohio, August 22.

Dubois, J.-M., S.K. Song, and Y. Massiani. 1993. Application of quasicrystalline alloys to surface coating of soft metals. Journal of Non-Crystalline Solids 153-154:443–445.

Elshafiey, I., and J. Gray. 1996. Optimization tool for x-ray radiography. Pp. 425–432 in Review of Progress in Quantitative Nondestructive Evaluation, Vol. 15. D.O. Thompson and D.E. Chimenti, eds. New York: Plenum Press.

Emeric, P.R., and W.P. Winfree. 1995. Thermal characterization of multilayer structures from transient thermal response. Pp. 475–482 in Review of Progress in Quantitative Nondestructive Evaluation, Vol. 14. D.O. Thompson and D.E. Chimenti, eds. New York: Plenum Press.

Erdogan, F., and K. Arin. 1972. A sandwich plate with a part-through and a debonding crack. Engineering Fracture Mechanics 4:449–458.

FAA (Federal Aviation Administration). 1996. Airworthiness Assurance R&D Branch: Research Accomplishments, 1996. Atlantic City, N.J.: FAA Technical Center.

Favro, L.D., X. Han, Y. Wang, P.K. Kuo, and R.L. Thomas. 1995. Pulsed echo thermal wave imaging. Pp 425–429 in Review of Progress in Quantitative Nondestructive Evaluation, Vol. 14. D.O. Thompson and D.E. Chimenti, eds. New York: Plenum Press.

Favro, L.D., X. Han, T. Ahmed, P.K. Kuo, and R.L. Thomas. 1996. Measuring corrosion thinning by thermal wave imaging. Pp. 374–379 in SPIE Proceedings, Vol. 2945. NDE of Aging Aircraft, Airports, and Aerospace Hardware. R.D. Rempt and A. Broz, eds. Bellingham, Washington: Society of Photo-Optical Instrumentation Engineers.

Fitzpatrick, G.L., D.D. Thome, R.L. Skaugset, and W.C.K. Shih. 1996. Magneto-optic eddy current imaging of subsurface corrosion and fatigue cracks in aging aircraft. Pp. 1159–1165 in Review of Progress in Quantitative Nondestructive Evaluation, Vol. 15. D.O. Thompson and D.E. Chimenti, eds. New York: Plenum Press.

Fredell, R.S. 1994. Damage tolerant repair techniques for pressurized aircraft fuselages. Ph.D. dissertation, Aerospace Engineering, Delft University of Technology, The Netherlands.

Fredell, R., C. Guijt, D. Conley, and S. Knighton. 1996. Design development of a bonded fuselage repair for the C-5A. Pp. 887–902 in Proceedings of the 1995 USAF Aircraft Structural Integrity Program Conference, Vol. 2. WL-TR-96-4094. Wright-Patterson AFB, Ohio: Wright Aeronautical Laboratories.

Geng, Z.J., G.G. Pan, W.S. Haynes, B.K. Wada, and J.A. Gorba. 1994. Six degree of freedom active vibration isolation and suppression experiments. Pp. 285–294 in Proceedings of the Fifth International Conference on Adaptive Structures. J. Tani, ed. Basel, Switzerland: Technomic.

Goering, J., and J. Dominguez. 1992. Large Scale Structural Damage Repair Analysis. Presentation at the 24th International SAMPE (Society for the Advancement of Materials and Process Engineering) Technical Conference, Toronto, Canada, October 20–22.

Goranson, U.G. 1997. Structural Airworthiness Initiatives. Presentation at the DoD/FAA/NASA Conference on Aging Aircraft, Ogden, Utah, July 8–10.

Gray, T.A., and R.B. Thompson. 1986. Use of models to predict ultrasonic NDE reliability. Pp. 911–918 in Review of Progress in Quantitative Nondestructive Evaluation, Vol. 5. D.O. Thompson and D.E. Chimenti, eds. New York: Plenum Press.

Hagemaier, D., and A. Hoggard. 1993. NDI technology as it relates to aging aircraft. Materials Evaluation 51(12):1360–1368.

Hagemaier, D., and G. Kach. 1997. Eddy current detection of short cracks under installed fasteners. Materials Evaluation 55(1):25–30.

Harris, C.E., J.H. Starnes, Jr., and J.C. Newman, Jr. 1995. Development of advanced structural analysis methodologies for predicting widespread fatigue damage in aircraft structure. Pp. 139–164 in FAA-NASA Sixth International Conference on the Continued Airworthiness of Aircraft Structures, C.A. Bigelow, ed. DOT/FAA/AF-95-86. Atlantic City, N.J.: FAA Technical Center.

Hart-Smith, L.J. 1974: Analysis and Design of Advanced Composite Bonded Joints. NASA-CR-2218. Washington, D.C.: National Aernautics and Space Administration.

Hegedus, C.R., S.J. Spadafora, and A.T. Eng. 1995. Organic coating technology for the protection of aircraft against corrosion. Pp. 17-1–17-10 in Corrosion Detection and Management of Advanced Airframe Materials. AGARD-CP-565. Advisory Group for Aerospace Research and Development. Neuilly-Sur Seine, France: North Atlantic Treaty Organization.

Hidano, L.A., and U.G. Goranson. 1995. Inspection programs for damage tolerance—Meeting the regulatory challenge. Pp. 193–211 in Proceedings of the FAA-NASA Sixth International Conference on the Continued Airworthiness of Aircraft Structures. DOT/FAA/AR-95/86. Atlantic City, N.J.: FAA Technical Center.

Hong, C.S., and K.Y. Jeng. 1985. Stress intensity factors in anisotropic sandwich plate with a part-through crack under mixed mode deformation. Engineering Fracture Mechanics 21:285–292.

Howard, M., and G. Mitchell. 1997. Evaluation of Wing Skin Fastener Hidden Corrosion Damage Assessment Technologies. Presentation at the DOD/FAA/NASA Conference on Aging Aircraft, Ogden, Utah, July 8–10.

Hsu, D.K., and T.C. Patton. 1993. Development of ultrasonic inspection for adhesive bonds in aging aircraft. Materials Evaluation 51(12):1390–1397.

JACG (Joint Aeronautical Commanders Group). 1996. U.S. Aviation S&T Roadmap: Aviation Vision, Vol. 1. Washington, D.C.: JACG S&T Process Board.

Jeans, L.L., G.C. Grimes, and H.-P. Kan. 1980. Fatigue Spectrum Sensitivity Study of Advanced Composite Materials. AFWAL-TR-80-3130. Wright-Patterson AFB, Ohio: Wright Aeronautical Laboratories.

Jones, R., and R.J. Callinan. 1979. Finite element analysis of patched cracks. Journal of Structural Mechanics 7(2):107–130.

Kan, H.-P. and M.M. Ratwani. 1981. Nonlinear behavior effects in cracked metal-to-composite bonded structures. Engineering Fracture Mechanics 15(1–2):123–130.

Kim, J., and N. Stubbs. 1995. Damage localization accuracy as a function of model uncertainty in the I-40 bridge over the Rio Grande. Paper 2446-23 in SPIE Proceedings, Vol. 2446. Smart Structures and Materials 1995: Smart Systems for Bridges, Structures, and Highways. L.K. Matthews and N. Stubbs, eds. Bellingham, Washington: Society of Photo-Optical Instrumentation Engineers.

Ko, W.L. 1978. An orthotropic sandwich plate containing a part-through crack under mixed mode deformation. Engineering Fracture Mechanics 10:15–23.

Komsky, I.N., and J.D. Achenbach. 1996. Ultrasonic imaging of corrosion and fatigue cracks in multilayered airplane structures. Pp. 380–388 in SPIE Proceedings, Vol. 2945. NDE of Aging Aircraft, Airports, and Aerospace Hardware. R.D. Rempt and A. Broz, eds. Bellingham, Washington: Society of Photo-Optical Instrumentation Engineers.

Komsky, I.N., J.D. Achenbach, G. Andrew, B. Grills, J. Register, G. Linkert, G.M. Huerto, A.L. Steinberg, M. Ashbaugh, D.G. Moore, G. Moore, and H. Weber. 1995. An ultrasonic technique to detect corrosion in DC-9 wing box from concept to field application. Materials Evaluation 53(7):848–852.

Leidheiser, H., and N. Das. 1975. Penetration of hydrogen into Al exposed to water. Journal of the Electrochemical Society 122(5):640–641.

Lichtenwalner, P.F., J.P. Dunne, R.S. Becker, and E.W. Baumann. 1997. Active damage interrogation system for structural health monitoring. Paper 3044-19 in SPIE Proceedings, Vol. 3044. Bellingham, Washington: Society of Photo-Optical Instrumentation Engineers.

Lincoln, J.W. 1996. Aging aircraft issues in the United States Air Force. SAMPE Journal 32(5):27–33.

Lincoln, J.W. 1997. Risk Assessments of Aging Aircraft. Presentation at the DOD/FAA/NASA Conference on Aging Aircraft, Ogden, Utah, July 8–10.

Ma, X.P., and J.P. Wikswo. 1996. Depth selective eddy current techniques for second layer flaw detection. Pp. 401–408 in Review of Progress in Quantitative Nondestructive Evaluation, Vol. 15. D.O. Thompson and D.E. Chimenti, eds. New York: Plenum Press.

Miller, R.N. 1987. Predictive Corrosion Modeling Phase I / Task II Summary Report. AFWAL-TR-4069. Wright-Patterson AFB, Ohio: Wright Aeronautical Laboratories.

Mindlin, H., B.F. Gilp, L.S. Elliott, M. Chamberlain, and T. Lynch. 1996. Corrosion in DOD Systems: Data Collection and Analysis (Phase I). MIAC Report 8. West Lafayette, Indiana: Metals Information Analysis Center (Purdue University).

Mitchell, R.A., R.M. Wooley, and D.J. Chivirut. 1975. Analysis of composite reinforced cutouts and cracks. AIAA Journal 13:774–749.

Moore, D. 1997. Navy Research in Aging Aircraft Technology. Presentation to the Committee on Aging of U.S. Air Force Aircraft, Washington, D.C., April 28.

Moulder, J.C., M.W. Kubovich, E. Uzal, and J.H. Rose. 1995. Pulsed eddy current measurements of corrosion-induced metal loss: Theory and experiment. Pp. 2065–2072 in

Review of Progress in Quantitative Nondestructive Evaluation, Vol. 14. D.O. Thompson and D.E. Chimenti, eds. New York: Plenum Press.

Moulder, J.C., J.A. Bieber, W.W. Ward, and J.H. Rose. 1996. Scanned pulsed eddy current instrument for nondestructive inspection of aging aircraft. Pp. 2–13 in SPIE Proceedings, Vol. 2945. NDE of Aging Aircraft, Airports, and Aerospace Hardware. R.D. Rempt and A. Broz, eds. Bellingham, Washington: Society of Photo-Optical Instrumentation Engineers.

Nakagawa, N. 1988. Eddy current detection methods for surface-breaking tight cracks. Pp. 245–250 in Review of Progress in Quantitative Nondestructive Evaluation, Vol. 7. D.O. Thompson and D.E. Chimenti, eds. New York: Plenum Press.

Nakagawa, N., and R.E. Beissner. 1990. Probability of tight crack detection via eddy current inspection. Pp. 893–899 in Review of Progress in Quantitative Nondestructive Evaluation, Vol. 9. D.O. Thompson and D.E. Chimenti, eds. New York: Plenum Press.

NRC (National Research Council). 1995. Expanding the Vision of Sensor Materials. NMAB-470. National Materials Advisory Board. Washington, D.C.: National Academy Press.

NRC. 1996a. New Materials for Next-Generation Commercial Transports. NMAB-476. National Materials Advisory Board. Washington, D.C.: National Academy Press.

NRC. 1996b. Accelerated Aging of Materials and Structures: The Effects of Long-Term Elevated Temperature Exposure. NMAB-479. National Materials Advisory Board. Washington, D.C.: National Academy Press.

NRC. 1997. Aging of U.S. Air Force Aircraft: Interim Report. Report No. NMAB 488-1. National Materials Advisory Board. Washington, D.C.: National Academy Press.

NTSB (National Transportation Safety Board). 1988. Aircraft Accident Report: Aloha Airlines, Flight 243, Boeing 737-200, N73711, Near Maui, Hawaii, April 28, 1988. NTSB/AAR-89/03. Washington, D.C.: NTSB.

Panhuise, V.E. 1989. Introduction. Pp. 663–665 in Metals Handbook, Volume 17: Nondestructive Evaluation and Quality Control. Metals Park, Ohio: ASM International.

Park, J.H., T. Ogiso, and S.N. Atluri. 1992. Analysis of cracks in aging aircraft structures, with and without composite-patch repairs. Computational Mechanics 10:169–201.

Podney, W., and J. Moulder. 1997. Electromagnetic microscope for deep pulsed eddy current inspections. Pp. 1037–1044 in Review of Progress in Quantitative Nondestructive Evaluation, Vol. 16. D.O. Thompson and D.E. Chimenti, eds. New York: Plenum Press.

Ratwani, M.M. 1996. Repair/refurbishment of military aircraft. Pp. 6-1–6-21 in Aging Combat Aircraft Fleets—Long Term Applications. AGARD-LS-206. Advisory Group for Aerospace Research and Development. Neuilly-Sur Seine, France: North Atlantic Treaty Organization.

Ricker, R.E., and D.J. Duquette. 1988. The role of hydrogen in corrosion fatigue of high purity Al-Zn-Mg exposed to water vapor. Metalurgical Transactions A 19A:1775–1783.

Ritchie, R.O., and J. Lankford. 1986. Small Fatigue Cracks. Warrendale, Pennsylvania: The Metallurgical Society of AIME.

Rogers, L., J. Maly, I. Searle, R. Ikegami, W. Owen, R. Gordon, and D. Conley. 1997. Durability Patch: Repair and Life Extension of High Cycle Fatigue Damage on Secondary Structure of Aging Aircraft. Presentation at the DoD/FAA/NASA Conference on Aging Aircraft, Ogden, Utah, July 8–10.

Rose, L.R.R. 1981. An application of inclusion analogy. International Journal on Solids and Structures 17:827–838.

Rudd, J.L. 1996. USAF aging aircraft program. Pp. 1-1–1-13 in Aging Combat Aircraft Fleets—Long Term Applications. Advisory Group for Aerospace Research and Development (AGARD) Lecture Series 206. Neuilly-Sur Seine, France: North Atlantic Treaty Organization.

Rummel, W.D. 1989. Applications of NDE reliability to systems. Pp. 674–688 in Metals Handbook, Vol. 17. Nondestructive Evaluation and Quality Control. Metals Park, Ohio: ASM International.

SAB (Scientific Advisory Board). 1994. Report of the Ad Hoc Committee on Life Extension and Mission Enhancement for Air Force Aircraft, Vol. 1. Executive Summary. U.S. Air Force Scientific Advisory Board Report No. SAB-TR-94-01. Washington, D.C.: Department of the Air Force.

SAB. 1996. Report of the Materials Degradation Panel, Ad Hoc Committee on Life Extension and Mission Enhancement for Air Force Aircraft, U.S. Air Force Scientific Advisory Board. Washington, D.C.: Department of the Air Force.

Saff, C.R., and M.A. Ferman. 1986. Fatigue life analysis of fuel tank skins under combined loads. Pp. 277–290 in Case Histories Involving Fatigue and Fracture Mechanics. ASTM-STP-918. Philadelphia, Pennsylvania: American Society for Testing and Materials.

Schmerr, L.W., and D.O. Thompson. 1994. NDE and design—a unified life-cycle engineering approach. Pp. 2183–2190 in Review of Progress in Quantitative Nondestructive Evaluation, Vol. 13. D.O. Thompson and D.E. Chimenti, eds. New York: Plenum Press.

Sethuraman, R., and S.K. Mathi. 1989. Determination of mixed mode stress intensity factors for a crack-stiffened panel. Engineering Fracture Mechanics 33:355–369.

Smith, S.W., and J.R. Scully. 1996. Hydrogen trapping and its correlation to the hydrogen embrittlement susceptibility of Al-Li-Cu-Zr alloys. Pp. 131–141 in Hydrogen Effects in Materials. A.W. Thompson and N.R. Moody, eds. Warrendale, Pennsylvania: The Minerals, Metals, and Materials Society.

REFERENCES

Spencer, F., and D. Schurman. 1995. Reliability Assessment of Airline Inspection Facilities: Vol. 3. Results of an Eddy Current Inspection Reliability Experiment. DOT/FAA/CT-92/12. Atlantic City, N.J.: FAA Technical Center.

SPIE (Society of Photo-Optical Instrumentation Engineers). 1996. Proceedings of NDE of Aging Aircraft, Airports, and Aerospace Hardware, Vol. 2945. R.D. Rempt and A. Broz, eds. Bellingham, Washington: SPIE.

Spiedel, M.O. 1975. Stress corrosion cracking of aluminum alloys. Metallurgical Transactions A 6A:631–651.

Sprowls, D.O., R.J. Bucci, B.M. Ponchel, R.L. Brazill, and P.E. Bretz. 1984. A Study of Environmental Characterization of Conventional and Advanced Aluminum Alloys for Selection and Design. Phase II—The Breaking Load Test Method. NASA-CR-172387. Washington, D.C.: National Aeronautics and Space Administration.

Syed, H.I., W.P. Winfree, K.E. Cramer, and D.A. Howell. 1995. Thermographic detection of corrosion in aircraft skins. Pp. 2035–2041 in Review of Progress in Quantitative Nondestructive Evaluation, Vol. 14. D.O. Thompson and D.E. Chimenti, eds. New York: Plenum Press.

Tarn, J.Q., and K.L. Shek. 1991. Analysis of crack plates with a bonded patch. Engineering Fracture Mechanics 40:1055–1065.

Taylor, S.R., N.P. Cella, G.E. Stoner, R.G. Buchheit, and L.P. Montes. 1997. Environmentally Compliant Corrosion Resistant and Electrically Conductive Inorganic Coatings for Aluminum Alloys. DARPA Progress Report, contract no. F49620-96-0305. Washington, D.C.: Defense Advanced Research Projects Agency.

Thome, D.K., G.L. Fitzpatrick, R.L. Skaugset, and W.C.L. Shih. 1996. Aircraft corrosion and crack inspection using advanced magneto-optic imaging technology. Pp. 365–373 in SPIE Proceedings, Vol. 2945. NDE of Aging Aircraft, Airports, and Aerospace Hardware. Rempt and A. Broz, eds. Bellingham, Washington: Society of Photo-Optical Instrumentation Engineers.

Thompson, D.O., and L.W. Schmerr, Jr. 1993. The role of modeling in the deterimination of probability of detection. Pp. 285–301 in Advances in Signal Processing for Nondestructive Evaluation of Materials. X.P.V. Moldague, ed. Boston: Kluwer Academic Publishers.

Ting, J., T. Jensen, and J. Gray. 1993. Using energy dispersive x-ray measurements for quantitative determination of materials loss due to corrosion. Pp. 1963–1969 in Review of Progress in Quantitative Nondestructive Evaluation, Vol. 12. D.O. Thompson and D.E. Chimenti, eds. New York: Plenum Press.

Vosteen, L.F., and R.N. Hadcock. 1994. Composite Chronicles: A Study of the Lessons Learned in the Development, Production, and Service of Composite Structures. NASA-CR-4620. Washington, D.C.: National Aeronautics and Space Administration.

Walter, P. 1995. The FAA's aging aircraft nondestructive inspection validation center at Sandia National Laboratories: An introduction. Materials Evaluation 53(7):799–802.

Whitehead, R.S., H. Kan, R. Cordero, and E. Saether. 1986. Certification Testing Methodology for Composite Structures. Final Report, March 1984–December 1985. Warminster, Pennsylvania: Naval Air Development Center.

Wincheski, B., J.R. Fulton, S. Nath, M. Namburg, and J.W. Simpson. 1994. Self-nulling eddy current probe for surface and subsurface flaw detection. Materials Evaluation 52(1):22–26.

Wincheski, B., R. Todhunter, and J. Simpson. 1997. A new instrument for the detection of cracks under airframe rivets. Pp. 2113–2120 in Review of Progress in Quantitative Nondestructive Evaluation, Vol. 16. D.O. Thompson and D.E. Chimenti, eds. New York: Plenum Press.

Winfree, W.P. 1996. New nondestructive techniques for inspection of aircraft structures. Pp. 2–13 in SPIE Proceedings, Vol. 2945. NDE of Aging Aircraft, Airports, and Aerospace Hardware. R.D. Rempt and A. Broz, eds. Bellingham, Washington: Society of Photo-Optical Instrumentation Engineers.

Xu, J., R. Wallingford, T. Jensen, and J. Gray. 1994. Recent developments in the x-ray simulation code: XRSIM. Pp. 557–562 in Review of Progress in Quantitative Nondestructive Evaluation, Vol. 13. D.O. Thompson and D.E. Chimenti, eds. New York: Plenum Press.

APPENDICES

Appendix A

Synopses of Air Force Aging Aircraft Structural Histories

This appendix provides a brief synopsis of the structural history for each of the Air Force's aging aircraft listed in Table 3-1. Also included are brief summaries of recent structural problems encountered on aircraft that have been reported to the committee by representatives of the system program directors, the Aeronautical Systems Center's engineering and technical management organization (ASC/EN), and the aircraft manufacturers.

AIR MOBILITY COMMAND AIRLIFTER AND TANKER AIRCRAFT

C/KC-135

The Air Force acquired the KC-135 tanker aircraft to replace the KC-97 to fulfill the need to refuel the B-52 bomber force. Boeing developed the prototype of this aircraft, designated the 367-80 or simply the Dash-80, with their own funds. The first flight of the Dash-80 took place in July 1954. From this aircraft, Boeing developed the KC-135, as well as the 707 and 720 commercial jet transport aircraft. The Air Force ordered limited production of the KC-135 in August 1954, and the first flight occurred in August 1956. Production continued until 1965, with a total of 820 aircraft manufactured. Thirty-seven different designations of the -135 aircraft have existed in the Air Force inventory. The active KC-135 force is still in excess of 600 aircraft with more than 550 of them in the tanker force. As of October 1994, the average use was 13,536 flight hours or 3,153 flights for the KC-135 and 28,361 flight hours and 3,108 flights for the RC-135.

To minimize structural weight and thus maximize payload capability, the Air Force elected to use 7178-T6 aluminum in the lower wing skins as well as in other locations in the aircraft along with 7075-T6 aluminum. The commercial 707 used 2024-T3 aluminum in the lower wing skins at about two-thirds the stress level. Other structural differences between the -135 and the 707 are found in the fuselage structure. For example, sections of the -135 lower fuselage (below the floor) that contain body fuel cells are not pressurized. Also, the -135 fuselage does not contain tear straps and shear ties between the fuselage frames and skins. Although the 707 had a design life goal of 20,000 flights or 60,000 flight hours, the Air Force did not specify a design service life goal for the KC-135.

In 1962 the Air Force decided to perform a full-scale fatigue test of the KC-135 to quantify its expected life. The test resulted in a failure of the wing at 55,000 simulated flight hours. A tear-down inspection of the failed wing revealed several hundred smaller cracks. From this test the Air Force estimated that the safe life of the wing was about 13,000 flight hours. However, by the late 1960s it became apparent that the wings were cracking earlier than expected and that the 13,000 hours, even if correct, probably was not sufficient to cover the projected future use of the aircraft. As a result, a second more realistic full-scale fatigue test was performed in 1972. This test eliminated the load excursions to 90 percent of limit load that were applied every 200 flights in the 1962 test. These load excursions were intended leave marker bands on fatigue crack surfaces to assist post-test evaluation, but actually artificially prolonged the test life by retarding the crack growth. More wing cracking occurred in the 1972 test than in the 1962 test, and complete failure occurred at 43,200 simulated flight hours. The existence of many small cracks early in the fatigue test raised a concern about the possibility of widespread fatigue damage (WFD) in service aircraft. Adding to the concern was the very small critical crack sizes in the 7178-T6 wing skin and the fact that by the mid-1970s there already had been a number of cases of unstable crack propagation and panel failures in service aircraft. Although the wing was fail-safe for these failures, the concern was that this fail-safety would be lost if WFD was present. Boeing performed tear-down inspections of six wings removed from service to determine the actual state of fatigue cracking in the wings of service aircraft. In 1977 the Air Force formed a blue ribbon panel to look at this problem. The panel concluded that the onset of WFD was occurring between 8,000 and 9,000 flight hours and recommended that (1) the lower surfaces of the wings be redesigned and replaced using the 2024-T3 aluminum used on the 707 and (2) flight restrictions be imposed on all aircraft that had already reached 8,500 flight hours. This led to a modification program to replace the center and inner lower surfaces of the wing out to wing station 733 production joint (i.e., just outside the outboard engines). The Air Force elected not to modify the 7178-T6 outer wing panels, but instead decided to cold work the fastener holes to enhance their resistance to fatigue cracking.

In 1977 the Air Force initiated a durability and damage tolerance assessment (DADTA) of the KC-135 with primary emphasis on the fuselage and empennage because the earlier modification had removed much of the concern about the wings. From this assessment, combined with an evaluation of the 1972 fatigue test results, it was estimated that, with the defined inspections and modifications, the aircraft could be flown safely well into the twenty-first century (i.e., it was estimated that the economical service life would extend to or beyond 2040 with the estimated utilization rates).

During the 1980s the reengine of the KC-135 aircraft was initiated to increase the aircraft's fuel off-load capability and reduce the noise levels. The old J57 engines were replaced with new CFM-56 engines on the active Air Force tankers (then redesignated KC-135R) and used JT-3D fan engines from retired 707 commercial aircraft on the Air National Guard tankers (then redesignated KC-135E). These engine modifications had little or no effect on the aircraft's structural integrity, even though some structural modifications were required.

Concern about structural deterioration due to corrosion led to a tear-down inspection of a retired KC-135 aircraft in 1991 by the Oklahoma City Air Logistics Center (ALC). This aircraft had spent 29 years at Mildenhall Air Base in the United Kingdom, which has a very corrosive environment. This inspection, and continuing corrosion investigations, provided considerable insight into the extent of corrosion and where to look, particularly the hidden areas. It also served as a testbed for evaluating various nondestructive inspection techniques. Corrosion and stress corrosion cracking (SCC) remain the primary structural issues concerning the KC-135. Specific concerns were that the life of the airframe would not meet the 2040 goal that was estimated by the 1977 DADTA because corrosion would accelerate the onset of WFD in the fuselage or empennage or make some of the calculated inspection intervals unconservative.

In 1995 the C/KC-135 system program director chartered the C/KC-135 aging aircraft integrated product team, known as Coral Reach, to develop an aircraft sustainment master plan, which defined a number of activities intended to enhance flight safety, reduce the cost per flying hour, and improve aircraft availability. In early 1996 this plan was reviewed by a blue ribbon panel consisting of representatives from the Air Force, NASA, and the FAA, with advisors from Boeing, the prime weapon system contractor. This panel concluded that from the available data it does not appear that there will be an onset of WFD in the fuselage, the empennage, or the previously modified lower wing surfaces until after 2040, assuming that the aircraft utilization rates and use remain as predicted. This conclusion was based partially on the results of tear-down inspections of fuselage panels removed from high-time 707/JSTARS aircraft. These panels were heavily corroded, but there was no evidence that this was causing early fatigue cracking. However, the panel acknowledged that there are differences between the 707 and KC-135 fuselage construction in some areas, and additional analytical investigations and inspections were needed to improve the estimates of the onset of WFD. Also, the panel expressed concern about the long-term effectiveness of the cold-worked fastener holes in the 7178-T6 aluminum lower surfaces of the outer wing (i.e., this stemmed from the fact that some small fatigue cracks had been reported in some outer wing fastener holes in one specific aircraft that had been inspected) and recommended several actions, including assessing the need to replace these lower wing surfaces and the station 733 joint closure rib that has had a lot of problems with corrosion and SCC. It is the committees understanding that the Air Force is now seriously considering this option. The 1996 blue ribbon panel also complimented Oklahoma City ALC on its maintenance program and emphasized the need to maintain aggressive efforts to prevent corrosion and SCC from becoming safety issues.

During the course of this study, the following specific corrosion and SCC problems were reported to the committee by a representative of the system program director:

- corrosion between fuselage lap joints and spot-welded doubler layers
- corrosion around fasteners in the 7178-T6 aluminum upper wing skins
- corrosion between wing skins and spars
- corrosion between bottom wing skin and main landing gear trunnion
- corrosion between fuselage skin and steel doublers around pilot windows
- SCC of large 7075-T6 aluminum forgings (fuselage station 620, 820, and 960)
- corrosion and SCC of fuselage station 880 and 890 floor beams
- corrosion and SCC of the wing station 733 closure rib
- corrosion in the E model engine struts

C-141B

The C-141A was designed and manufactured by Lockheed (now Lockheed-Martin) as a long-range, heavy logistics transport aircraft. The primary materials in the aircraft are the 7000-series aluminum alloys heat treated to the T6 condition. A total of 285 aircraft were manufactured and delivered to the Air Force from January 1964 to February 1968. The original design life goal for the aircraft was 30,000 flight hours, and a full-scale fatigue test was performed to validate this design goal. In addition, the aircraft was designed to be fail-safe for a single-element failure (e.g., a single wing plank), which was then the standard for commercial aircraft design. The Aircraft Structural Integrity Program (ASIP) also included an individual aircraft tracking program (IATP).

By 1974, after the aircraft had been in service for about 10 years, it was evident that the fuselage was volume limited for a number of logistics missions. A decision was made to add approximately 22 ft. to the length of the fuselage and to add in-flight refueling capability to the aircraft. However, before the Air Force was willing to expend the funds on this effort, it wanted to know if there was enough remaining life to justify the modifications. The Air Force's Aeronautical Systems Division (now Aeronautical Systems Center) recommended that an update of a 1975/1976 C-141A DADTA be performed to determine this justification and to define additional modification and inspection requirements. This assessment was performed in 1977 and early 1978 and concluded that the lower-bound economic service life was 45,000 hours of the then-current use spectrum (called the SLA-II spectrum). The aircraft fuselages were extended and the aircraft were redesignated as the C-141B.

By late 1992 the aircraft had reached an average of about 35,000 equivalent SLA-II spectrum hours, with some higher-time aircraft approaching the 45,000-hour economic service life estimate. Also, by then the aircraft had been experiencing many fatigue cracking and corrosion problems. Because of delays and uncertainty about the future of the C-17, which was to replace the C-141, Congress, in their FY93 authorization bill, directed the Secretary of the Air Force to convene a Scientific Advisory Board (SAB) committee to determine the technical feasibility of extending the service life of the C-141. This committee was convened early in 1993, held a series of meetings during the first half of that year, and released a final report in January 1994.

At the time of the SAB committee reviews there was increasing evidence of the onset of WFD in several different locations in the wings, corrosion and SCC in the upper surface of the center wing, fatigue cracking and SCC around the windshield, fatigue cracking in the stiffeners in the aft pressure door, SCC in the fuselage main frames, and some corrosion in the empennage.

Tear-down inspection of the wings from two service aircraft, which had about 45,000 equivalent SLA-II spectrum hours, showed evidence of WFD in the fuel drain holes (i.e., weep holes) in the integral risers in the lower surfaces of the wings. Methods to protect the structural safety until aircraft retirement or lower-surface replacement by inspections, hole cold working, and the use of bonded composite doublers were being investigated. Also, WFD had been found previously in the wing station 405 chordwise joint that connects the inner wing to the outer wing, and a modification consisting of a large doubler plate plus many local repairs and hole oversizings was already under way. Although the corrosion and SCC was a serious maintenance problem in the center wing box, the most serious concern from the standpoint of flight safety was the fatigue cracking that was occurring in the joggle area of the lower-surface side-of-body chordwise joint (wing station 77). At the time of the review, 72 aircraft had had their center wing boxes refurbished and this lower joint reinforced. The structural safety of the remainder of the aircraft was being protected by frequent close inspections until they could be modified or the aircraft retired. The final area of the wing that was a concern form the standpoint of WFD, and probably the most difficult in that there was no identified modification or repair short of lower-surface replacement, was the spanwise splices that connect the multiple wing panels together. During the 1977/1978 DADTA and again in a review in 1990, it had been predicted that the onset of WFD in these splices would occur at about 45,000 equivalent hours of the SLA-II spectrum. The tear-down inspection of the wings that had about 45,000 hours had revealed some cracking, but the inspections were not complete and no final judgment about the adequacy of the 45,000-hour limit was made by the SAB committee. None of the fuselage or empennage cracking and corrosion problems were considered to be life limiting by the SAB committee, and various modifications and repairs were under way. However, the SAB committee identified several areas of the aircraft where corrosion was causing major economic problems.

Since the 1993 SAB committee review the weep hole cracking problem was brought successfully under control through a combination of inspections, the use of bonded boron/epoxy doublers, and, where possible, cold working of holes. This took a concerted effort by the Air Force's Wright Laboratories, Warner-Robins ALC, and their supporting contractors. Also, the modifications to the wing station 405 splice were completed, and inspections, modifications, and repairs in the other areas of the aircraft continue to take place. With regard to the onset of WFD in the wing spanwise splices, there have been additional inspections in operational aircraft and more cracking has been found. Using these findings, Lockheed-Martin has performed a risk analysis and has concluded that the previous 45,000-hour estimate for the onset of WFD is unconservative. They now believe that 37,000 equivalent hours of SLA-II is a better estimate of the onset, causing concern over the fail-safety of all aircraft with a greater number of hours. The only alternative to grounding (or replacing the lower wing surface) is to protect the structural safety through frequent, careful, and very burdensome inspections of all highly-stressed fastener holes in the spanwise splices to detect and repair cracks before they reach critical size. This will require the inspection of over 6,000 fastener holes per aircraft every 120 days until the aircraft is retired. The C-141Bs are now in the process of being retired and replaced by the C-17, but as seen in Figure 2-4, they will not be completely phased out of the inventory for several more years. Until it is retired, the structural management of this force will continue to be a significant challenge.

C-5

Lockheed was awarded the contract for the C-5A airlifter in October 1965. The first flight was in June 1968, and by the end of 1970 thirty aircraft out of a total production buy of eighty-one had already flown and were delivered to the Military Airlift Command. All of the production aircraft had been delivered by May 1973. Full-scale static and fatigue testing that was conducted in 1971 revealed serious structural deficiencies (discussed below) that led to a major wing modification program called H-Mod. No changes or modifications were incorporated during production on any of the 81 aircraft because of the lateness of the testing[1] and because of the type of contract that the government had with Lockheed (i.e., fixed price under a total package procurement concept). The Air Force authorized Lockheed to proceed with H-Mod in 1978, and by 1987 all surviving C-5As had been modified. Also, in 1982 the decision was made to have Lockheed build 50 C-5Bs, which incorporated the wing improvements included in the H-Mod. This program was completed in 1988. As of September 1996 there were a total of 126 C-5s (As and Bs) in the inventory. This included 81 in the active force, 32 in the Air Force Reserve, and 13 in the Air National Guard. The average age of the total C-5 force is 18 years; however, the average age of the A model is about 26 years. The flight-hour distributions for the C-5A and C-5B aircraft are shown in Figure A-1.

The C-5 structure is a multiple-load-path fail-safe design that is made predominately of 7000-series aluminum alloys. The wings consist of a center wing box, inner wing boxes, and outer wing boxes. The upper and lower surfaces of the boxes are made of shiplap planks fastened together by single rows of interference fit fasteners. The individual planks are machined from thick aluminum plate with integral spanwise stiffeners. The planks in the original C-5 wings were made from 7075-T6 aluminum, but in H-Mod they were changed to 7075-T73511. The original contract for the C-5 specified the maximum empty weight as a design requirement, but provided for the 30,000-hour design life as a noncontractual design goal. Thus, when difficulties were encountered in achieving the weight requirement, the contractor elected to reduce the thickness of the wing planks, which raised the operating stress levels in the wings. The resulting stress levels were significantly higher than the contractor had used previously in the design of the C-141 wings and they were much higher (e.g., 40 to 80 percent higher) than those used on any commercial transport aircraft. It was hoped that the adverse effect of these higher stresses on structural life would be offset by improved quality (i.e., lower stress concentration factors) and the beneficial effects of the interference-fit fasteners, and thus the design life goal would still be achieved. This did not turn out to be the case.

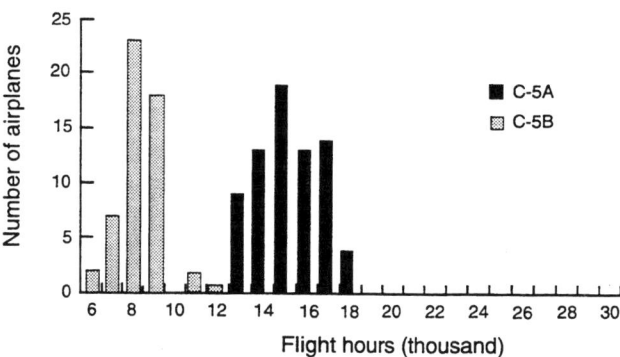

FIGURE A-1 C-5 flying hour distribution (through March 1996).

The first major structural difficulty occurred in the spring of 1971 when there was a tension failure of the lower surface of the wing at less than the design ultimate strength during the full-scale static test program. Next there was a series of local failure and cracking problems very early during full-scale fatigue testing. The three major problem areas were (1) the chordwise joints at wing station 120 that connect the inner wings to the center wing box, (2) the spar webs, and (3) the spanwise splices that connect the shiplap wing planks together. In fact, it appeared that the wings were in a state of WFD after less than one-third of a design lifetime of cyclic testing. At this point the Air Force and Lockheed jointly decided that an independent review team (IRT), consisting of government and industry aircraft structures experts from outside the C-5 program, should be formed to perform an in-depth review of the C-5 structural design (with primary emphasis on the wings) and determine available solutions. This effort was conducted in 1972 and was one of the first DADTAs of Air Force aircraft.

Although the wings were designed to withstand the loss of a single wing plank without a complete wing failure, the IRT had two concerns about the fail-safe design. First, there was a concern that if a plank failed due to manufacturing damage (e.g., a rogue flaw) in a spanwise splice fastener hole, there was a high probability that the flaw would exist in two planks (i.e., due to the common fastener hole) and as a result there would be a two-plank failure that the wing could not sustain. Second, even if only one plank failed, the wing would not be fail-safe if the structure reached the point where there was WFD (i.e., there were many small cracks in the adjacent wing planks). Thus, the IRT determined safety limits for the C-5 wings based on slow crack growth from an assumed maximum probable initial manufacturing damage in a spanwise splice fastener hole and also by estimating the onset of WFD based on the fatigue test results. In addition, they developed and evaluated numerous potential near- and long-term solutions, including load alleviation options, several types of fastener changes, various local reinforcements, and redesign of major portions of the wings. The IRT recommended fuel

---

[1] This led to the ASIP requirement that one lifetime of full-scale fatigue testing be completed before full production authorization on future weapon systems.

management and a load alleviation system to reduce the near-term fatigue damage rate and development of an essentially new, lower-stressed wing (i.e., plan H or H-Mod) for incorporation before reaching the estimated safety limit of the original wings. They recommended that the use of the aircraft then in operation be limited to 6,500 hours. This was based on the 1972 use spectrum (i.e., a 14-mission spectrum) and assumed that a passive load distribution system was implemented. This limit was based on safe crack growth from an initial manufacturing flaw in a spanwise splice fastener hole. The estimated time for onset of WFD was only slightly higher at about 8,000 hours of the same spectrum.

Following the IRT study, the ongoing C-5 program at Lockheed continued to evaluate the options, refine the H-Mod design, review the use of the operational aircraft, refine the damage tolerance calculations, and further evaluate the findings from the fatigue test articles. Also, Air Force Headquarters had asked the RAND Corporation to perform an airlift study, which included another independent look at the C-5 structures problem. There was considerable resistance against the Air Force immediately committing to a major modification program for the C-5A because of the costs involved, uncertainty over the extent of modifications that were really needed, and when it was needed, and the perception that Lockheed would be rewarded for fixing a problem that many felt was of their own making. In January 1975, the Air Force Aeronautical Systems Division (ASD) asked a committee of the SAB's Division Advisory Group to review the new data and results of the analyses and evaluations performed since the 1972 IRT and assess their potential impact on the need for H-Mod and recommendations for when it should be initiated.

By the time of the January 1975 review by the Division Advisory Group committee, the operational spectrum for the C-5A had been changed from the original 14-mission spectrum to a new spectrum called the "representative mission profiles," or the RMP spectrum, so as to more nearly reflect the actual and planned aircraft use. Also, the tear-down inspection of the full-scale fatigue test wings had revealed more spar web cracking than had previously been thought to exist, and the damage tolerance analyses had been refined to include the effects of shear loading on crack growth rates. The additional spar web cracking indicated that new spar webs were likely required for H-Mod, rather than only repairs as had been previously thought. The net result of the changes in the spectrum and analysis method resulted in a predicted safety limit of 8,000 RMP hours based on slow crack growth from an assumed 0.05-in. initial manufacturing flaw in a spanwise splice fastener hole to critical size. It was also estimated that the onset of widespread fatigue cracking may be as high as 10,000 RMP hours. The Division Advisory Group committee reiterated the need for H-Mod and further recommended an active load distribution system rather than the passive system then in use. This would further increase the safety limit by about 10 percent.

The position taken by the RAND Corporation from their study was that the Lockheed and Air Force analyses were too conservative and that the onset of widespread fatigue cracking may be in the 12,000 to 15,000-RMP-hour range. Also, they were skeptical about the need for a modification as extensive as H-Mod and believed that it would outlast the rest of the aircraft. They further recommended information enhancement initiatives to better define the C-5's structural modification needs. The ASD agreed that there were uncertainties in the safety limit prediction and the estimate of the onset of WFD, but did not agree that they were overly conservative. ASD also supported the recommendation for information enhancement initiatives, specifically further residual strength analysis and tests, nondestructive evaluation (NDE) development, risk analysis, and a phased tear-down inspection of the wings from a high-time aircraft to determine if widespread cracking had initiated. These recommendations led to the 1977 Structural Information Enhancement Program (SIEP), which was, in effect, another DADTA of the C-5A wing structure.

As part of the 1977 SIEP effort, the slow crack growth safety limit was refined further, based on additional analyses and tests, to be 7,100 RMP hours compared with the 8,000-RMP-hour value at the time of the 1975 Division Advisory Group committee review. Also, as part of this SIEP effort, a wing that had been on an operational aircraft (Lockheed no. 68-0214) was torn down and inspected. This aircraft had accumulated about 6,700 RMP hours at the time of tear down. A total of 44,641 fastener holes were inspected, and 1,361 small cracks were detected. Of these, 931 were considered significant. Although initial manufacturing damage was noted in some holes, none was as large as was assumed in the safety limit calculation. On the other hand, the number of small cracks was more than anticipated. A risk assessment was performed to determine whether or not the C-5A wings would have lost their fail-safety given a single-plank failure from any cause (e.g., impact from an engine burst or gunfire), assuming that the crack population found in aircraft 68-0214 was representative of the other aircraft. The results of this analysis predicted that the failure probability was about 2×10^{-3} at the time the 7,100-RMP-hour safety limit is reached. Although this was higher than the 1×10^{-4} that had been established previously as criteria for the onset of WFD, the Air Force group monitoring the SIEP activities did not recommend a reduction in the 7,100-hour safety limit and scheduled time for H-Mod. However, an enhanced special inspection program was recommended on each aircraft until H-Mod was accomplished. During the 1980s the H-Mod program proceeded relatively free from further disruption until it was completed in 1987.[2]

---

[2] Congressional hearings were held during 1980 by the Joint Committee on Economics chaired by Senator William Proxmire. These hearings investigated the circumstances surrounding the award of the H-Mod contract to Lockheed.

In November 1996 this committee received a briefing from a representative from the San Antonio ALC on the current structural problems encountered on the C-5 aircraft. Not surprisingly, there was no mention of any fatigue cracking problems in either the C-5A H-Mod wings or the C-5B wings because they now have low operating stress levels. However, it was somewhat surprising that, given the complexity of the structure, there was no mention of any fatigue cracking problems in the fuselage or the empennage. The dominant structural problems encountered to date have been SCC of the 7075-T6 aluminum mainframes, keelbeam, and fittings in the fuselage; and SCC of the 7079-T6 fuselage lower lobe and aft upper crown.

AIR COMBAT COMMAND BOMBER, FIGHTER, AND ATTACK AIRCRAFT

B-52H

The B-52H was the last model of the B-52 strategic bombers built by Boeing. A total of 102 B-52Hs were built during 1961 and 1962. Of these, the Air Force still has 85 in the active force and 9 in the Air Force Reserve. The average flight hours of these aircraft was about 13,500 in 1995. The high-time aircraft had 18,313 hours and 2,363 flights.

Structurally, the B-52H was nearly the same as the B-52G, and both models underwent major structural modifications during the 1960s when the Air Force changed the mission of the aircraft from a high-level strategic bomber to a low-level penetrator. These modifications incorporated some tougher materials (e.g., the lower wing skins were changed from 7178-T6 to 2024-T3 aluminum, and the in-board upper wing skins were changed from 7178-T6 to 7075-T6 aluminum), lower stresses, and improved structural details. The extent of the three largest modifications (i.e., ECPs 1050, 1128, and 1185) is illustrated in Figure A-2. The design life goal of this modified wing and body structure was 12,000 flight hours, and it was fatigue tested to 72,000 cyclic test hours or six lifetimes during the 1960s. After the test, the tear-down inspection revealed about 222 cracks. In early 1978 a structures working group led by ASD/EN conducted a review of the B-52G/H structures to obtain an estimate of the longevity of the airframe because several expensive upgrades of the avionics systems and weapons carriage were planned. This group concluded that the results of the 1960s full-scale fatigue test may have been somewhat optimistic because the test contained periodic overloads that would artificially retard crack growth. The test spectrum was severely truncated, and the severity of the spectrum was not well defined compared with actual service use. Also, even after the 1960s modification programs, there was still a lot of 7178-T6 and 7075-T6

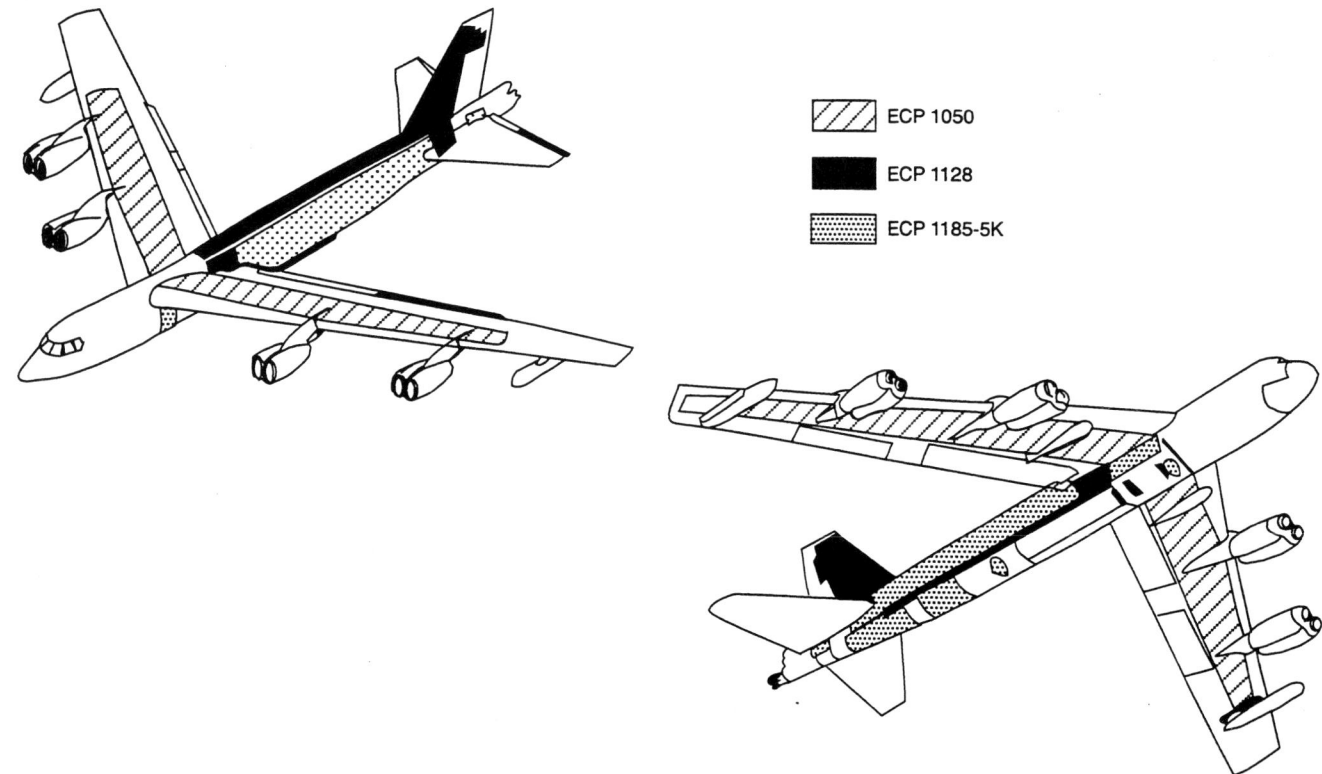

FIGURE A-2 General locations for B-52G/H structural improvements.

aluminum in the primary structure of the airframe. These alloys have relatively poor fracture properties and are susceptible to corrosion and SCC. The group recommended that a detailed DADTA be conducted. This was performed during 1979 and early 1980 and the tracking program was upgraded. The DADTA identified about two dozen critical areas requiring inspections and potential future modification. It also concluded that, with inspections, modifications, and continued tracking of the aircraft, the aircraft could be operated safely into the twenty-first century.

In 1990 a B-52 structures working group was formed with representatives from the Oklahoma City ALC and engineers from Boeing with expertise in fatigue, stress, design, and materials. The purpose of this group was to develop solutions to current structural problems and to address long-term aging issues. Some specific problems they have addressed in the past several years are

- cracking in the bulkhead at body station 694
- fatigue cracking in flap tracks
- cracking in the side skin of the pressure cabin
- cracking in aft body skins
- cracking in the upper surface of the wing
- fatigue cracking in the thrust brace lug of the forward engine support bulkhead

Force-wide inspections and identified corrective actions for each of these problems have either been or are being implemented. In addition, a tear-down inspection of a retired service aircraft was conducted to assess the corrosion problems on the aircraft, and an updated DADTA was completed in 1995. The corrosion tear-down inspection revealed only relatively minor problems, and it was concluded that, with continued use of corrosion-preventative compounds, corrosion should not be an issue. As part of the DADTA, an estimate was made of the lower-bound economic life of the airframe. It was determined that the limiting component was the upper surface of the wing, where it was determined that the lower-bound economic limit was about 32,000 hours of the current use spectrum. Based on the current utilization rate, it was estimated that this would allow the aircraft to be used beyond the year 2030. This is illustrated in Figure A-3.

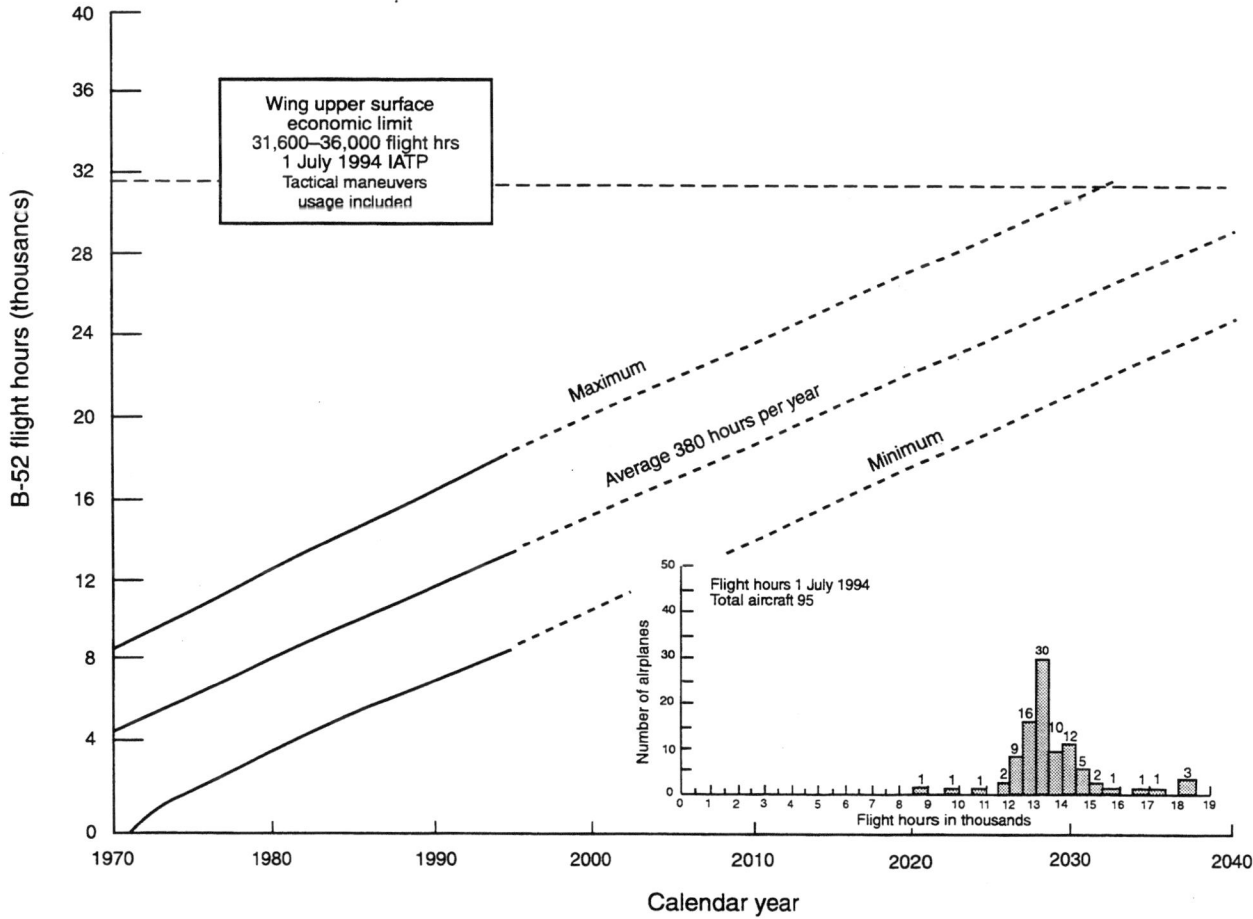

FIGURE A-3 B-52H current use rate.

B-1B

The North American Rockwell B-1 strategic bomber program was awarded in June 1970. The program was to consist of 4 prototypes and 244 production aircraft. The first prototype flew in December 1974, and the program was terminated in 1977 in favor of cruise missiles. A 100-aircraft program was resurrected in October 1981 as the B-1B. The contract was awarded in January 1982 and the first flight of the B-1B was in October 1984 and the initial operational capability (IOC) was in October 1986. Currently, there are 81 B-1Bs in the Air Force and 14 in the Air National Guard.

Although the initial contract did not have damage tolerance design requirements, design changes were made in 1971 to implement them. This represented the first Air Force aircraft designed to such requirements. Although the requirements had not yet matured to the level reflected in MIL-STD-1530A and MIL-A-83444, which were released in 1975, the requirements did include a safe crack growth design that significantly influenced the materials selection. The material selected for the wings was 2219 aluminum, which had been used successfully in the Apollo space program. Titanium 6A1-4V was used for the wing carry-through structure and the internal structure of the horizontal tail, and a tough low-carbon steel alloy was used for much of the empennage support structure. Considerable effort went into obtaining fracture toughness data and fatigue crack growth data for all of the alloys used in the B-1. It was these data that formed the start of the Air Force *Damage Tolerance Handbook*.

Designed to be damage tolerant, the in-service structural inspection requirements would be expected to be minimal if the aircraft were flown close to their original design loads spectrum. However, as with many other combat aircraft, the B-1B actual use spectrum is more severe than the original design spectrum. The load factor occurrences have been in excess of the design use, particularly in traffic patterns. In fact, there have been occurrences of load factors in excess of the design limits for the aircraft. Also, the fuel reserves at landing are in excess of those assumed during the design. The reason for this is the desire to be able to avoid commercial airports in the event of the need to use an alternate airport. The consequence of the increased load factor occurrences and increased landing weights is increased fatigue damage rate (i.e., the rate of flaw growth) and thus shortened inspection intervals. This could become a problem in the wing structure if fastener removal becomes necessary for inspection because the B-1B uses interference fit fasteners in the wing. This is currently being investigated by the Oklahoma City ALC to determine the alternatives. Obviously, the best solution would be to return to the original use for which the aircraft was designed, if possible, since the aircraft are still quite young and the damage to date may not be excessive.

To date, the fatigue cracking and corrosion problems on the B-1B airframes appear to be minimal and, unlike many earlier aircraft, more emphasis was placed on selecting materials with increased corrosion resistance. One exception has been the horizontal tails that have encountered high-cycle fatigue damage to the internal titanium sine wave spar structure. The tail is located just above the exhaust wake of the engines, but well within the high-acoustic-noise envelope. This placement of the tail was intentional to achieve high-performance turns at low velocities using the engine exhaust to increase the control power exercised by the horizontal tail. Although the original ground fatigue tests and ground vibration tests showed no problems and the natural frequencies of the tail were beyond those at which acoustically driven problems would be expected, flight experience showed that high-cycle fatigue cracking occurred very early in the service life. After considerable effort on the part of the contractor, the problem was found to be caused by the fact that the production tails had gaps between the skins and the titanium substructure. This caused bending of the spar and rib flanges during assembly, producing high sustained stresses in the flanges. In addition, there was loosening of the blind fasteners in flight. These factors caused the response frequency of the overall tail to be reduced such that it fell within excitation frequencies of the engine acoustic noise. The contractor developed an overall analysis of the dynamic response of the tail, which in turn has led to a modification that is believed to have solved this problem.

F-15

The McDonnell Douglas F-15 was the winner of an Air Force competition for a new air superiority fighter aircraft in December 1968. The first flight of the aircraft was in July 1972 and IOC was in January 1976. Since the start of the program, five different models have been built for the Air Force (plus models for foreign military sales). These are the single-seat A and C models, the two-seat B and D models, and the dual-role two-seat F-15E Strike Eagle. The Air National Guard currently has 116 of the A and B models, and the Air Force has 621 of the other models. The structural configuration of the aging A through D models are much the same.

The F-15A/D models were designed under Air Force ASIP requirements in the late 1960s prior to the adoption of damage tolerance requirements; however, full-scale fatigue and static testing were also conducted. A complete DADTA was performed in the early 1980s to update the maintenance program based on the damage tolerance approach. In the initial design of the F-15, McDonnell Douglas incorporated a fatigue-resistant interference fastener system, but ignored its beneficial effects when establishing the operating stress levels for the structure. This turned out to be fortuitous in that it allowed some margin for increase in severity of the loads spectrum. In fact, the growth in weight of the aircraft and changes in load factor severity have significantly increased the spectrum severity, causing the time required to grow an initial flaw to critical size to be reduced to about one-fourth of its original value. This increased severity was noted through the IATP conducted by the Warner-Robins ALC, and because the change was so significant, it was decided to conduct an additional full-scale fatigue test to the increased severity spectrum. This test was conducted at the Wright Laboratories test facility at Wright-Patterson AFB. The results of this test indicate that the original operational service life goal of 8,000 hours should still be attainable. However, the increased use severity will increase the inspection burden, and some of the wing inspections could become particularly onerous because of the current lack of NDE capability to inspect for small cracks without removal of the fasteners. McDonnell Douglas is currently using the results of the tear-down inspection of the fatigue test aircraft and crack growth analyses to obtain a better estimate of the actual service life expectancy of the F-15.

When the F-15E was designed, the MIL-STD-1520 and MIL-A-83444 damage tolerance requirements had been implemented. This meant that some areas of the original F-15 structures design had to be changed to meet these requirements, and some additional testing was required to prove the structural integrity. To date, the E models seems to be flying close to their design use spectrum.

The structural problems that have been encountered in service on the F-15 fall into the five following general categories:

- damage to honeycomb structure
- buffet-induced cracking
- acoustic-induced cracking
- corrosion in nonhoneycomb structure
- low-cycle fatigue cracking

The F-15C and E models have experienced honeycomb water intrusion, corrosion, disbonds, cracks, and in-flight loss of various secondary structures such as wing tips, ailerons, flaps, fin leading edges, and horizontal tail components. These problems have been caused by leak paths, inadequate bond durability, and unexpected dynamic loading. The current solution has been to perform a patch repair or to replace the components with improved honeycomb components.

The areas of the F-15 structure that have encountered buffet-induced cracking are illustrated in Figure A-4. Twin-tailed aircraft, such as the F-15 and the Navy's F-18, use vortices generated from the fuselage at high angles of attack to provide additional rudder power for control. Unfortunately, these same vortices provide a very turbulent flow field at intermediate angles of attack and subject the tails to a high-frequency, asymmetric loading that causes early high-cycle fatigue cracking and partial failure of the tail structure. The first sign of cracking due to these loads in the F-15 was in the pod attachments at the top of the tails. Local repairs did nothing but move the failure points and reduce the life. It took a careful analysis of the entire tail response and fuselage attachment stiffness by McDonnell Douglas to simulate the tail vibration modes and deflections that led to these failures and provide the insight to arrive at a solution. This involved increasing the overall stiffness of the tail by adding graphite composite plies to attenuate the vibration. In the case of wing cracking due to buffet, as indicated in Figure A-4, the cause was flow detachment over the outer wing at even modest angles of attack that resulted in high-frequency out-of-plane loading. These loads vibrated the skins and integral stiffeners and caused cracking of the rib mouseholes through which the spanwise stiffeners ran. Again, local repairs did not solve the problem. Eventually, damping systems were applied to the stiffener/rib connections to reduce the problem, and an alternate method of connecting the stiffener cap to the rib was developed.

The primary acoustically induced high-cycle fatigue cracking on the F-15 was encountered on the E model after stores (externally mounted weapons and systems) were qualified for use on the aircraft. The E model is configured for both air-to-air and air-to-ground attack missions, and in the air-to-ground mission radar evasion often requires low-altitude, high-speed cruise and dash to the target. With multiple stores attached to pylons beneath the wings, shocks are formed, which cause high acoustic vibrations to occur on certain skin panels of the fuselage. These vibratory loads have been high enough to cause high-cycle fatigue cracking of some skin panels. To permanently fix such damage, it is necessary to design the repaired structure such that its natural frequency is out of the range of the shock impingement frequency. This is a complex problem that requires knowledge of both the excitation sources and the structural responses. The current approach to fixing these problems on the F-15 has been to replace the damaged structure with parts with greater thickness to increase strength and to apply damping material. Additional research in understanding and developing repairs or modifications for these types of problems (e.g., composite repairs and better damping materials) appears worthwhile.

The corrosion problems in nonhoneycomb structure on the F-15 have been minimal. There have been some problems in the fuselage fuel tank, the outboard leading-edge structure of the wings, and the flap hinge beam. The current solution has been to improve drainage, repair, and replace.

The primary low-cycle fatigue cracking that has occurred in service to date has been in the upper surface of the wing in compression-designed structure that was not sized for fatigue during the initial design. Also, there has been one fuselage cracking problem. The specific low-cycle fatigue cracking locations were as follows:

- upper wing surface stringer runouts
- upper wing spar cap seal grooves
- front wing spar conduit hole
- upper in-board longeron splice plate holes

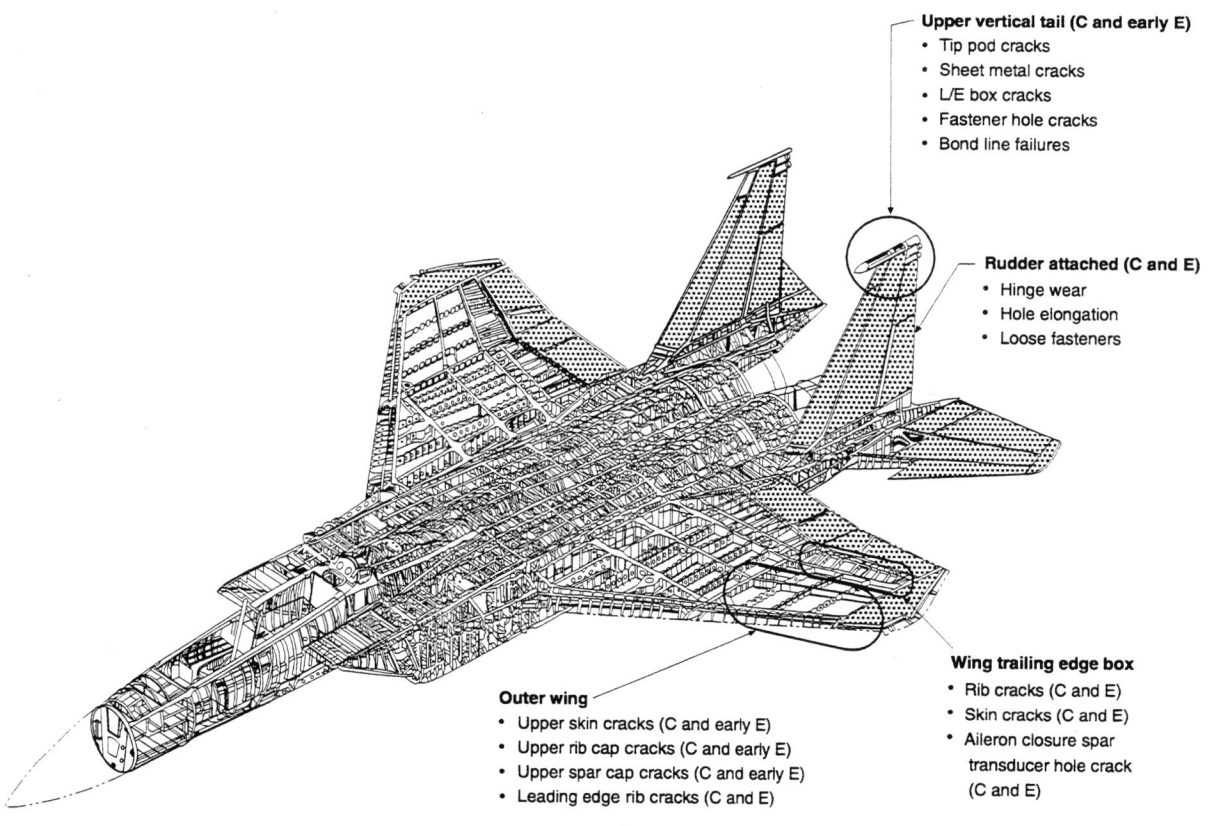

FIGURE A-4 F-15 buffet-induced problems.

None of these problems is life limiting, and in all cases preventive repairs have been designed and installed on the aircraft.

F-16

The first General Dynamics YF-16 prototype flew in January 1974. After winning the fly-off against the YF-17 prototype, eight more development aircraft (i.e., six single-seat A models and two two-seat B models) were built. The production authorization was announced in the spring of 1978. The first flight of a production F-16A was in August 1978 and IOC was in 1979 at the 388th Tactical Fighter Wing at Hill AFB. The Air Force currently has 809 F-16s in the active forces, 73 in the Air Force Reserve, and 631 in the Air National Guard.

When the full-scale development began in January 1975, the airframe was designed to the ASIP requirements in MIL-STD-1530A and the damage tolerance requirements in MIL-A-83444. Materials and stress levels were selected to preclude the need for structural inspections throughout the design operational life of 8,000 hours of the design use spectrum. The structural arrangement of the F-16 is shown in Figure A-5. The aircraft was designed for a limit load factor of 9 g's to what was thought to be a severe load factor exceedance curve. The design gross weight of the aircraft was 22,500 lb., and the design mission distribution was

- 55.5 percent air-to-air
- 20.0 percent air-to-ground
- 24.5 percent general

A full-scale static test was conducted under more than 100 different test loading conditions. All test objectives were met, and no structural modifications were needed. A full-scale fatigue (durability) test was conducted to two lifetimes of the design use spectrum and was completed in March 1978. Some cracking occurred in the fuselage center section bulkhead shear webs, and action was taken to reinforce these areas in the production aircraft. Also, a few other cracks occurred at local stress concentrations, which required some local changes in the production aircraft.

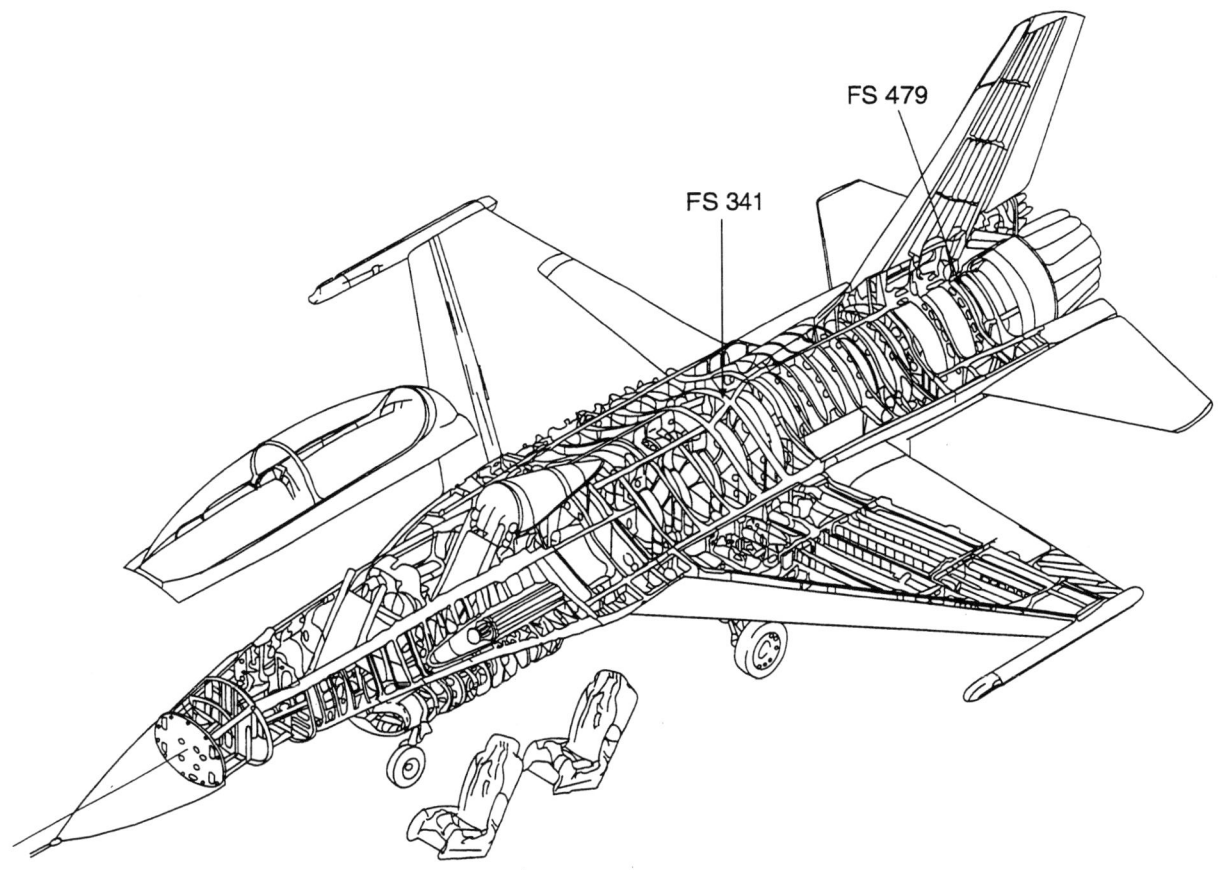

FIGURE A-5 F-16 structural arrangement.

Almost immediately after its introduction into operational use, the gross weight of the aircraft started to increase due to increased payloads, and the aircraft use changed. In 1981, General Dynamics made an assessment of the structural capability of the F-16 at a gross weight of 23,500 lbs. and concluded that the aircraft was adequate from the standpoint of static strength. The effect of the increased weight and mission changes on the fatigue was still uncertain because there was still only a limited amount of tail number tracking data available. By the mid-1980s the new mission distribution was

- 28 percent air-to-air
- 57 percent air-to-ground
- 15 percent general

Also, the data from the IATP began to indicate that the load factor (N_z) exceedances were more severe than had been assumed in the initial design for all three missions. It was believed that these increased exceedances were caused by an "Alpha-g" limiter that allowed the pilots to accomplish significantly more high-"g" maneuvers without fear of overloading the aircraft. This, combined with the continuing increases in aircraft weight (e.g., the design gross weight for the F-16C/D, block 50, grew to 28,750 lbs.), resulted in significantly increased stress exceedances and reduced fatigue life. In February 1984 the F-16 Systems Program Office asked the Aeronautical Systems Division's engineering and technical management organization (ASD/EN; now ASC/EN) to perform an independent assessment of the F-16 structural integrity program. Based on this review it was decided that the use severity dictated the need for a new full-scale static and durability test.

During the full-scale static test in October 1987, the left wing failed at approximately 85 percent of the design ultimate strength (as a result of the maximum wing bending moment being 25 percent higher for the F-16C/D than for the original F-16A/B). A modification was developed and the static test was successfully completed. The full-scale fatigue (durability) test was begun in September 1987. A primary purpose of this new test was to identify the areas of the structure that had become fatigue critical as a result of the more severe and different use of the aircraft. The test revealed about a dozen new critical areas by October 1989, when it had reached 7,330 cyclic test hours. At this point the test was stopped for replacement of the wing attach bulkheads (i.e., fatigue cracks were first noted in the bulkheads at about 4,000 cyclic test hours). Damage tolerance analyses were performed to establish the safety limits and inspection requirements for all of critical areas.

In early 1991, the Air Force, led by ASD/EN, conducted another independent review of the F-16 structural integrity program. By this time there were 18 critical areas or locations identified in the airframe. Of the 18 areas, 15 had been identified by either the 1987/1989 full-scale fatigue test or subsequent component testing. Of these 15, 6 had also shown up as cracking problems on service aircraft. In addition, two more areas had shown up on service aircraft that had not been identified by fatigue testing. One was at a tab radius of the wing attach bulkhead and the other was at a pad radius of the vertical tail attachment bulkhead at fuselage station 479. It was later discovered from the flight recorded data that the actual service loading spectrum for the vertical tail was more severe than that which had been applied during the 1987/1989 full-scale fatigue test because of rolling maneuvers with rudder input from the pilot. One critical area was identified by analysis only. The review team concluded that the potential for future service problems was high, and they made a number of recommendations with regard to future inspections and modifications.

By mid-1995 more service cracking had been discovered, generally at times fairly close to what had occurred during fatigue testing, and repair and modification plans were either in place or being defined for the various models (i.e., A, B, C, and D) and blocks of aircraft within a model. The six general areas requiring repair or modification are shown in Figure A-6, along with the ECP number that addressed the F-16C/D block 40 modifications.

In November 1996 the committee received a briefing by the system program director's representative on current structural problems on the F-16. The six problem areas that were highlighted were

- cracking of the vertical tail attachment bulkhead at fuselage station 479
- cracking of fuel vent holes of the lower wing skin
- cracking of the wing attach bulkhead at fuselage station 341
- cracking of the upper wing skin
- fastener problems on the horizontal tail support box-beam
- cracking of the ventral fin

In each case, the current repair and replacement concepts were described. Concerns about future fatigue cracking were expressed, along with the possibility of hidden corrosion. However, corrosion was not discussed as being a current problem.

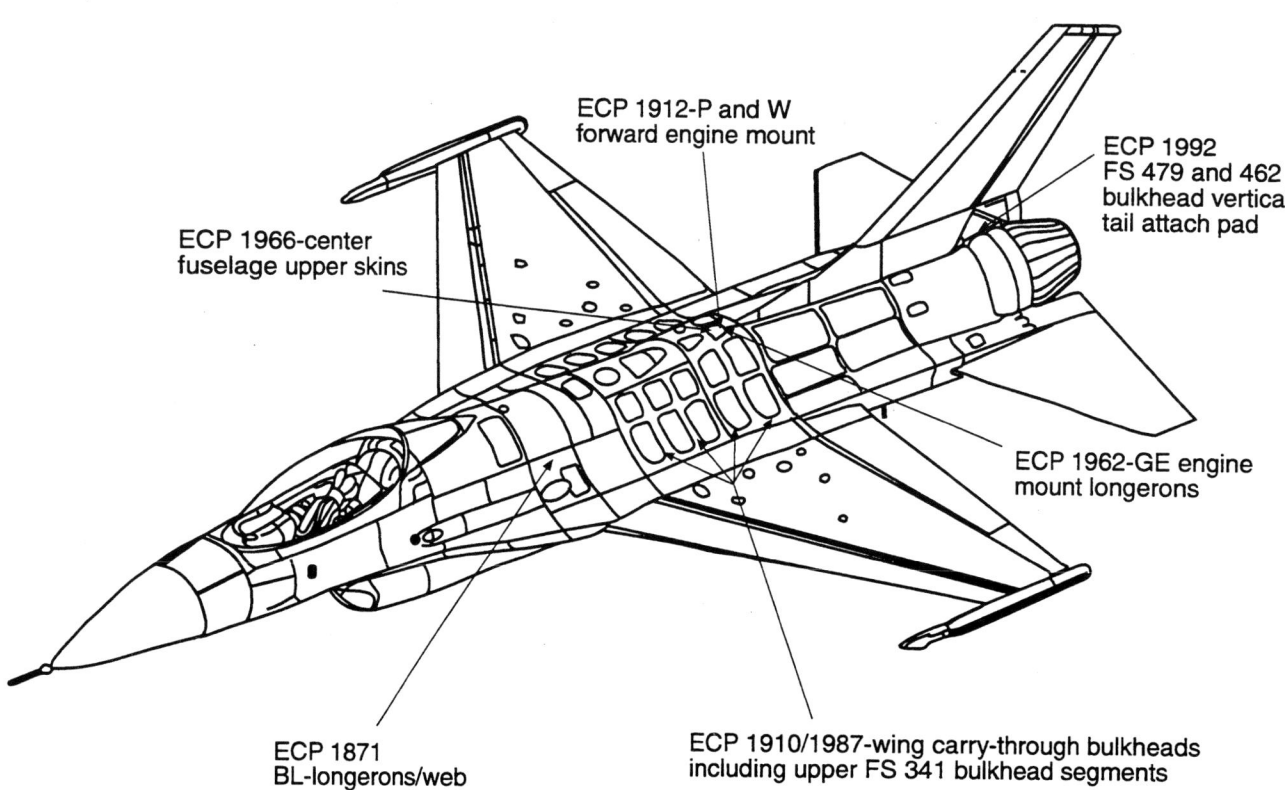

FIGURE A-6 F-16 structural modification areas.

A-10

The Fairchild-Republic A-10 was selected by the Air Force over Northrop's competitive A-9 prototype in January 1973. This was followed by full-scale development, and production began in 1975. The IOC was in 1977 and the first units were deployed to Europe in 1978. The Air Force currently has 223 A/OA-10s in the active force, 51 in the Air Force Reserve, and 101 in the Air National Guard. As of April 1996 the flight hours on the aircraft ranged from about 3,500 to about 6,800, with an average of about 5,000.

The structural configuration of the A-10 is shown in Figure A-7. The original design life for the A-10 was 6,000 hours to a design use spectrum that was developed largely from data that had been required from previous ground attack aircraft such as the A-7 and F-4. Also, the Air Force ASD was convinced by the operators and the contractor that only mission-required fuel should be used in the development of loading spectra for the aircraft. Both of these factors turned out to be serious errors. First, the design spectrum failed to recognize that this aircraft, when operating at low levels, would be subjected to many more evasive maneuvers than had been experienced previously by either the A-7 or the F-4. This resulted in a higher number of load exceedances than the aircraft was designed for. Second, the Tactical Air Command operated the aircraft with full fuel tanks rather than using only mission fuel as was originally intended. The combination of these two factors resulted in the A-10 actual use being about three times more severe than the original design use. Further compounding the problem was the recognition, early on, that the 6,000-hour design life was too low and that it should be increased to 8,000 hours. Although the goal of the original full-scale fatigue test was to achieve two lifetimes (i.e., 12,000 hours) of the original design spectrum (which it did obtain with some problems), it was necessary to extend the testing and increase the severity of the test spectrum. Thus, after testing to 12,000 hours of the original spectrum, the test was continued another 3,480 hours of the new severe spectrum (called spectrum 3) with repairs to the wing structure. In addition, it became apparent that there was a need to perform a full DADTA using the new spectrum to determine the safety limits and inspection

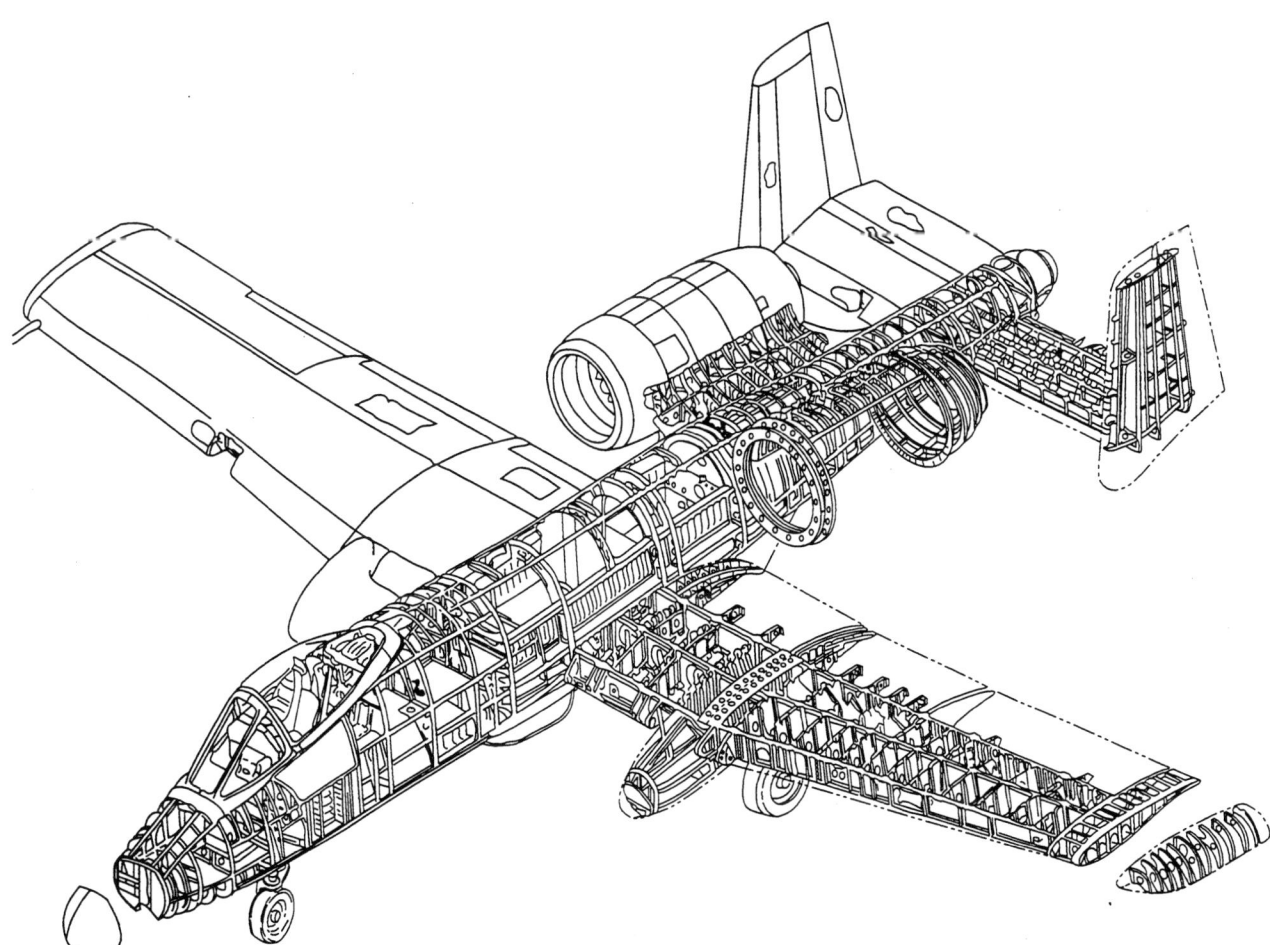

FIGURE A-7 A-10 structural arrangement.

requirements for all critical areas for both modified and unmodified structure. Accordingly, an on-site Air Force senior structures engineer was assigned at Fairchild-Republic from June 1978 to September 1979 to lead this effort. A total of 68 critical areas were identified and the inspection and modification needs defined. Additional full-scale wing, empennage, and forward fuselage component fatigue testing to the spectrum-3 loading was also accomplished.

The full-scale fatigue testing resulted in a number of cracking problems. In the center panel of the wing there was skin, spar cap, and spar web cracking at numerous fastener holes. As a result, up to 1,500 holes per wing were cold worked as a near-term maintenance measure. In addition, the structure was redesigned to incorporate thicker lower skins and spar caps to achieve the desired 8,000 hours of spectrum-3 use. There were also cracks in web of the front spars emanating from nutplate holes around access cutouts. In the outer wing panels, fatigue cracks from fastener holes resulted in complete wing fracture at wing station 135 and 178. As a result, the lower skin was redesigned with a thicker three-step skin. This change was incorporated in production and retrofit on operational aircraft. In addition, there was cracking of the webs of both the front and the mid spars. In the fuselage, cracking occurred at the upper auxiliary longeron splice strap at fuselage station 524, and the frame at fuselage station 405 failed during the test at approximately 82 percent of the design lifetime. Both areas required redesign and changes to the operational aircraft. A fatigue cracking problem requiring redesign was also encountered when testing the main landing gear.

In addition to the cracking that occurred during the fatigue testing, there have been a number of other fatigue cracking problems discovered in operational aircraft as a result of the in-service inspections that have been performed. These have occurred at the locations listed below:

Wing:
- auxiliary spar cutout of the center section rib at wing station 90
- outer panel front spar web at wing station 118 to 126
- outer panel upper skin at leading edge

Fuselage:
- center fuselage forward fuel cell floor at the boost pump
- forward fuselage gun bay compartment
- forward fuselage lower longeron and skin at fuselage station 254
- center fuselage overwing lower floor panel stiffeners

Nacelle:
- aft nacelle hanger frame
- aft nacelle thrust fitting
- engine inlet ring assembly skin/frame

Main landing gear:
- shock strut outer cylinder

In 1991 a follow-on or update to the 1978/1980 DADTA was performed. The number of critical areas or locations had grown to 103. This assessment utilized an updated spectrum based on 477,440 total hours of individual aircraft tracking and 5,895 hours of flight data recorder data. These data indicated that the actual use was still more severe than the original design use, but not as severe as spectrum 3 which was based on much less flight data. The safety limits and inspection requirements were adjusted accordingly.

During the November 1996 review with this committee, the representative of the A-10 system program director listed the current known fatigue cracking problems on the service aircraft. In addition to some of the service problems listed above, he noted that fatigue cracking was occurring in the forward fuselage upper crown skin and at the lower wing skin fastener holes and pylon stud holes at wing station 23. Also, areas of the airframe where corrosion has been found were described, including

- Exfoliation corrosion:
  - 2024-T351 aluminum lower wing skin (chemically milled step)
  - 7075-T6 aluminum upper wing at the leading edge
  - 2024-T3511 aluminum lower front spar cap
  - other local areas in lower wing skin
  - 7075-T6 aluminum fuselage bottom skin 2024-T3/7075-T6 aluminum fuselage side skin and beaded pan
  - 2024-T3511 aluminum horizontal stabilizer upper spar caps

- Pitting corrosion:
  - 9Ni-4Co-0.3C steel wing attach fitting bushing and lug bore
  - main landing gear fitting attach bolts
  - 7075-T6 aluminum aft fuel cell aft bulkhead
  - 2024-T351 center fuselage upper longeron

- SCC:
  - wing attach bushing flange
  - main landing gear attach bolts

OTHER AIRCRAFT OF THE AIR COMBAT COMMAND

E-3A (AWACS)

Boeing was awarded a contract for two prototype airborne warning and control system (AWACS) aircraft in July 1970. The first production aircraft designated the E-3A was delivered to the 552nd Airborne Warning and Control Wing at Tinker AFB, Oklahoma, in March 1977. The Air Force currently has 32 E-3s in the active inventory. As of September 1995, the flight hours on these aircraft ranged from 9,809 to 15,872, with an average of 13,994. The number of flights ranged from 1,358 to 2,391, with an average of 1,885.

The E-3 is a derivative of the Boeing commercial 707-320B aircraft. The primary structural modifications were made to the aft fuselage to attach the support struts for the large fiberglass rotodome assembly. Designed in the early 1950s, the 707 airframe contains many parts made from the corrosion-susceptible 7000-series aluminum alloys in the T6 condition; however, the lower wing skins and the fuselage skins are made from the tougher 2024-T3 aluminum. The basic structure was designed to be fail-safe and was certified to the required fail-safe residual strength requirement then in existence (i.e., to be able to carry 80 percent of limit load after failure of a structural member or a large partial failure). At that time there was no requirement for fatigue testing and none was performed on the wing. Hydro-fatigue testing and fail-safe testing was performed on the fuselage. The fail-safe testing consisted of dropping guillotine blades through large sections of the fuselage, and more than 30 such tests were conducted during the aircraft development.

In 1976 the Air Force contracted with Boeing to perform a DADTA on the E-3 to establish structural inspections based on the MIL-A-83444 damage tolerance requirements and to assess the probable durability of the airframe in the anticipated Air Force use. Comparative analyses between the commercial 707 and the E-3 were performed to allow the interpretation of the commercial 707 (lead-the-fleet) service experience in relation to future E-3 maintenance needs.

The original design life goal for the commercial 707 aircraft was 20,000 flights and 60,000 flight hours and, as can be seen in Table 3-3, there are 707s that have exceeded these goals that are still in commercial service. As noted above, the average E-3 has accumulated less than 10 percent of the 20,000-flight commercial design goal, and as such, the onset of WFD should not be a concern for a number of years.

During the review with this committee in November 1996, the representative from the E-3 program at the Oklahoma City ALC indicated that the structural service experience included some isolated fatigue damage and generalized corrosion of the 7000-series aluminum alloys and especially the 7178-T6. The specific problems noted were:

- Fatigue and corrosion:
  - rudder skins
  - spoiler actuator clevis

- Exfoliation corrosion:
  - 7XXX-T6 upper wing skin
  - leading-edge slats
  - main landing gear door
  - fillet flap
  - magnesium parts
  - fuselage stringer 23

- Delaminations and disbonds:
  - windows, floor panels, and nose radome core

- Wear:
  - antenna pedestal turntable bearings

E-8 (JSTARS)

The Northrop Grumman E-8 joint surveillance and attack radar system (JSTARS) program consists of two E-8A prototypes, one preproduction E-8B, and nineteen production E-8Cs. All of the E-8s are scheduled to be delivered to the 93rd Air Control Wing at Warner-Robins AFB, Georgia, by 2004, with the first production deliveries in 1997. The E-8 airframes are used Boeing 707 commercial aircraft. The flight hours on the first ten aircraft selected for production JSTARS range from about 40,000 to 64,000 hours, with an average of 53,615 hours. The number of flights on these aircraft range from about 17,200 to 22,250, with an average of 19,861. The planned future use for the aircraft is 16 years and 20,000 flight hours; however, this could very well be extended if the concept remains successful. The original design life goal for the commercial 707 aircraft was 20,000 flights and 60,000 flight hours.

During the refurbishment and modification of the initial commercial 707s to the E-8 configuration, corrosion was found to be quite extensive in the aircraft fuselage. Many longitudinal lap splices were opened up (i.e., the fasteners were removed) and the corroded materials were ground away. Where the corrosion was too severe, the skin panels and stringers were replaced. Fourteen complete panels and four partial panels removed from the first two production E-8s were sent to Boeing for detailed inspection. Both of the aircraft from which these panels were removed had seen about one design lifetime (i.e., approximately 20,000 flights and 60,000 hours) of commercial service and both had corrosion damage ranging from light (i.e., less than 0.001 in. deep) to severe (i.e., more than 0.01 in. deep). The purpose of the inspection was to determine if there was any indication of fatigue crack initiation, which could portend the future onset of WFD, and if corrosion was contributing to the crack initiation. Thousands of fastener holes, spot welds, and repair details were examined by a close visual inspection, followed by detailed inspections (under 20X magnification) of about 500 fastener holes that were in the more-severely corroded areas. Selected fastener holes were examined further using stereoscopic microscopes. No fatigue cracks were found in any of the fastener holes or other structural details. This finding, combined with the results of the original hydro-fatigue testing and the results of fatigue testing performed on the KC-135 (which is of similar construction in many areas), provided the Air Force with confidence that the E-8 fuselages (as well as the C-18 and the VC-137 fuselages that were also derived from the 707) will not experience the onset of WFD in the near future.

Extensive corrosion was also found around fasteners in the upper wing skins. The corrosion in these areas was ground out, many fastener holes were then cold worked, and many repairs were made in compliance with the Boeing repair manual for the 707 aircraft.

Because of previous evidence of some fatigue cracking in commercial 707 wings (i.e., primarily in the 7075-T6 aluminum stringers), the Air Force contracted with Boeing to perform detailed tear-down inspections of some sections of wings taken from two 707 aircraft that had been in storage at Davis Monthan AFB in Tucson, Arizona. The purpose of these inspections was to obtain data to predict when the 707 wings might be expected to experience the onset of WFD. One aircraft was a -300-series aircraft, such as the E-8s, that had 57,382 flight hours and 22,533 flights (approximately one design lifetime of commercial use). The other aircraft was a -100-series aircraft that had 78,416 flight hours and 36,359 flights. Figure A-8 shows the plan form of the 707-300 wing and the five sections that were removed for detailed inspection. These inspections revealed 1,084 small fatigue cracks in fastener holes in the 7075-T6 aluminum stringers and 591 small fatigue cracks in the fastener holes in the 2024-T3 aluminum skins of the 707-300 aircraft. The findings from the inspection of the 707-100 wing yielded somewhat fewer cracks (i.e., a total of 673 stringer and skin cracks were found) because a much smaller area of the wing was inspected. Ninety-six percent of the fatigue cracks that were found in the 707-300 were in the size range from 0.01 in. to 0.06 in., and 4 percent were greater than 0.06 in. long. The largest stringer crack completely severed one flange of the stringer, and the largest skin crack was < 0.20 in. out of each side of the fastener hole.

The questions that needed to be answered with regard to these tear-down inspection findings were:

1. Are the findings representative of the cracking that could be expected to exist in the other used commercial 707-300 aircraft that have had a similar number of flights or flight hours?
2. Are the cracks that were found of sufficient size and density to constitute the onset of WFD (i.e., are they of sufficient size and density to degrade the fail-safe residual strength of the wing to below the required level)?
3. If they have not yet reached the size and density necessary for the onset of WFD, when is it predicted that this will occur?

With regard to the first question, nothing was identified to indicate that the 707-300 aircraft that was torn down was not typical of any other 707 aircraft with similar use. Thus, it must

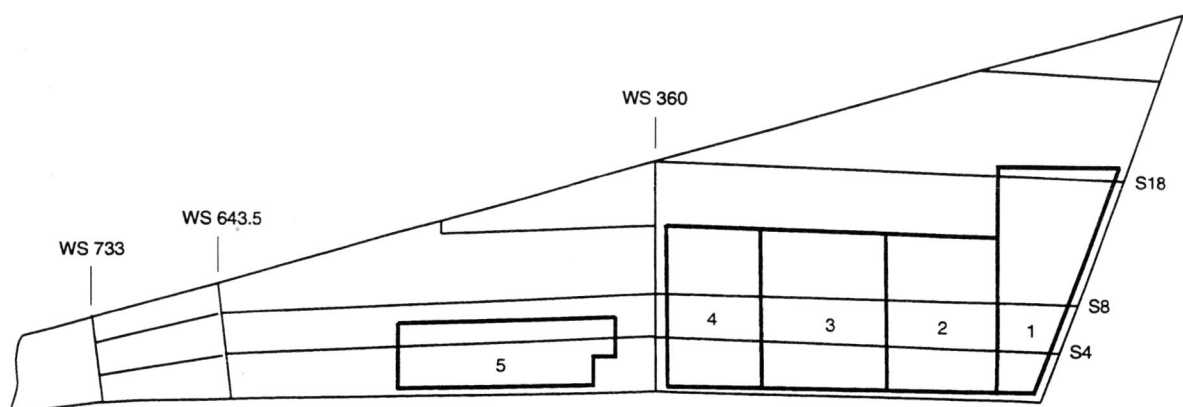

FIGURE A-8 Boeing 707 wing tear-down locations.

be assumed that similar fatigue cracking exists in the wings of all 707s, which have accumulated as many flights and flight hours.

The answer to the second and third questions depends on the fail-safe damage size and residual strength level that is required. The ASC/EN obtained the assistance of Boeing to analytically determine the residual strength of the wing, assuming the failure of two adjacent skin panels and the central stringer due to discrete source impact (e.g., impact from an engine burst or gunfire) or due to any other cause. The results of this analysis indicated that, with this damage size, the remaining wing structure was barely able to sustain the limit load under the assumption that there was no fatigue cracking in the adjacent stringers or skin panels. Although this damage size and residual strength exceed the original certification requirement for the 707 aircraft,[3] it is the typical design criteria for present day commercial aircraft. Also, ASC/EN considered it to be an appropriate fail-safe requirement for the E-8, recognizing that the E-8 could be operating in a much more hostile environment than the typical commercial 707 aircraft. Using the results of the residual strength analyses and the inspection findings from the tear down of the 707-300 aircraft, the ASC/EN, with the assistance of Boeing, developed crack distribution, crack growth, stress distribution, and critical stress functions necessary to perform a risk assessment of the E-8 wings. The results of this assessment indicated that, for the tear-down 707-300 aircraft, the risk of an aircraft loss—given the discrete source damage—was unacceptable beyond about 50,000 flight hours (i.e., the onset of WFD was at about 50,000 hours).

It is the committee's understanding that the Air Force is currently in the process of reviewing the options available to deal with this wing cracking problem in the E-8. For the lower-time aircraft, where the cracks are still very small, the option may be to remove the fasteners, clean the cracks out by oversizing the holes, and then cold expanding the holes to delay reinitiation. Where it is not possible to clean the cracks out by oversizing, the use of composite patches may be an option. For the higher-time aircraft the only option may be to replace the skin panels and stringers. It is also the committee's understanding that Boeing is assessing the implications of the tear-down inspection findings on the degradation of the fail-safety of the remaining commercial 707 aircraft and determining the need for corrective actions.

The findings also have obvious implications to the degradation in the fail-safety of the other 707 commercial-derivative aircraft that have accumulated a high number of flights and flight hours (i.e., some C-18 and VC-137 aircraft).

---

[3]The original design (in compliance with Civil Aviation Regulation 4b.270) was to be able to carry 80 percent of the limit load after failure of a single principal structural element or an obvious partial failure.

C-130

The Air Force selected the Lockheed proposal for what was then designated as a heavy cargo aircraft in 1951. After winning the competition, two YC-130 preproduction or prototype aircraft were designed and built during the period 1952 to 1954 with the first flight in August 1954. The first production contract was awarded in 1953 for seven C-130A models with the first flight in April 1955. The IOC was in 1956. The C-130s have five basic models: the A, B, E, H, and J. The current production model is the H with the future J model incorporating new engines and a two-man heads-up display cockpit. The Air Force currently has 311 C-130s in the active force, 141 in the Air Force Reserve, and 242 in the Air National Guard, for a total of 694 aircraft. Except for ten AC-130As still in the Special Operations Command, the C-130 force consists of E and H model aircraft. All of the B model aircraft were phased out of the Air Force's inventory by 1995. Nearly one-fourth of all the Air Force's C-130s are used in the various special purpose missions listed below.

- gunship (AC-130A/E/H/U)
- aerial drone launcher (DC-130E/H)
- electronic combat (EC-130E/H)
- search and rescue (HC-130H)
- helicopter tanker (HC-130N/P)
- ski airlifter (LC-130H)
- missile tracker and satellite recovery (JC-130H)
- multirole and special operations (MC-130E/H)
- weather reconnaissance (WC-130E/H)

The E model aircraft were delivered to the Air Force between 1961 and 1972, and the H models have been supplied since 1973. The average age of all the C-130s in the Air Force inventory is about 25 years.

Like most military aircraft designed in the 1950s and 1960s the C-130s used mostly 7000-series aluminum alloys heat treated to the T6 condition, and as a result they have encountered many corrosion and SCC problems over the years, in addition to many fatigue cracking problems. Some of the fatigue cracking problems are attributable to the large amount of low-level flying associated with the many special uses of the aircraft. The first serious problems occurred with corrosion and cracking of the center wing structure. This led to a redesigned center wing being incorporated in the production of the E model in 1968 and, in the period from 1968 to 1972 the center wings were replaced on all B models and the earlier E models.

Although there have been some minor improvements, the same center wing has been in production since the redesign in 1968. During the 1970s it was discovered that the damage tolerance of the outer wings was severely degraded due to fatigue cracking plus some faulty depot maintenance actions by some commercial contractors. In the 1979 to 1981 time period there was a miniDADTA conducted, which focused on this outer wing problem. During 1981 to 1983, a more complete DADTA was conducted. The results of these assessments led to the recommendation that the outer wing panels of all aircraft prior to Lockheed serial number 4542 be replaced with a new lower-stressed H model outer wing that was designed and put into production in 1984. During the period from 1984 to 1988, all of the outer wings on the Air Force C-130s built prior to Lockheed serial number 4542 were replaced. By 1993 fatigue cracking problems were again appearing in the center wings on certain versions of the C-130s assigned to the Special Operations Command (i.e., HC-130N/P, AC-130H, and MC-130E). This led to the initiation of a program to again replace the center wings on these aircraft with new center wings, which contain design improvements to accommodate the more severe use of this command.

During the 1992 summer study of the SAB, which addressed the technologies to support the Air Force's Global Reach/Global Power Concept, the board's Mobility Panel reviewed the structural status of the airlift aircraft. With regard to the C-130 they noted that the wings should not be a structural problem in the near future because of the many replacements that have been accomplished. On the other hand, they also noted that the average remaining life of the fuselage was quite low based on predictions made by Lockheed. Lockheed predicted an average remaining life for the E model fuselages of about 10,000 hours, with that of the high-time aircraft being considerably less. The aircraft were typically accumulating about 700 to 800 hours per year. In addition, the SAB pointed out that the validity of the estimate was questionable because of the lack of good fatigue test data (i.e., the full-scale fatigue test performed on the A model in 1956 consisted of only pressure testing to 20,000 cycles, and there had been structural changes since then). They recommended that the Air Force Matériel Command (AFMC) make a detailed review of the C-130 fuselage life estimate to assess if a new full-scale fatigue test was justified and what other life-extension measures were needed.

Although this committee saw no evidence to indicate that the Air Force acted on the SAB's 1992 recommendation, in March 1996 the Director of Logistics of the Air Combat Command wrote a memorandum to AFMC Headquarters requesting their assistance in updating the service life limit for the C-130 fuselages, since they were reaching the life limit projected by Lockheed. In response to this request, AFMC put together an ad hoc team, consisting of representatives from ASC/EN, Warner-Robins ALC, Wright Laboratories, and the FAA to investigate the issue. This ad hoc team identified five options for obtaining an improved estimate of the service life limit of the C-130 fuselage:

1. A full-scale durability test (i.e., a full-scale fatigue test), which simulated both internal pressure and external flight loads.
2. A detailed tear-down inspection of a high-time C-130 operational aircraft to identify the possible onset of WFD, critical areas, and corrosion problems.
3. A large panel fatigue test of the rear fuselage upper crown skin area.
4. A detailed DADTA of the fuselage.
5. A combination of options 2, 3, and 4.

The team recommended option 5 as the most cost effective. In addition, they recommended the continued enforcement of the C-130 corrosion tracking program and the use of current detection methods to search for corrosion and to develop corrosion signatures of the C-130s at each programmed depot maintenance to help determine trends in corrosion damage.

EF-111A

The EF-111A are F-111A aircraft produced by General Dynamics that have been converted to electronic countermeasures tactical jamming aircraft by Grumman. The original F-111 contract was awarded in 1962, and the first two prototype aircraft flew in 1964. The first production F-111A entered service in 1967. The F-111As that were later converted to EF-111As were built in the late 1960s. In 1975 Grumman received a contract to convert two F-111As to EF-111A prototypes. The first flight of an EF-111A with complete electronic systems was in 1977. After four years of operational testing, the first operational aircraft entered service in 1981. A total of 42 EF-111As were produced, with the last one delivered to the Air Force in 1985. There are currently 37 in the Air Force inventory. The average age of their airframes is about 29 years and the average flight hours is about 6,000. The current Air Force plan is to retire all of these aircraft over the next four or five years and have the Navy assume the mission using the Navy's EA-6B aircraft.

The F-111 airframe has a significant amount of high-strength D6ac steel in the wing carry-through structure, the tail support structure, and the fuselage longerons (see Figure A-9). Most of these components are heat treated to 220,000-psi tensile strength with some heat treated to the 260,000- to 280,000-psi range. The remainder of the airframe structure is fabricated mostly from aluminum alloys. The design load factors (N_z values) are -3 g to +7.33 g, and the original design life goals were 4,000 flight hours and ten years of service. At the time of the original design, the Air Force had not yet developed and implemented damage tolerance design requirements. The structure was designed with the so-called safe-life approach using a Miner's rule fatigue analysis. The operational life was limited to that demonstrated by the full-scale fatigue test reduced by a factor of four to account for data scatter.

A full-scale static test program was conducted, and after several local redesigns to correct strength deficiencies, the test

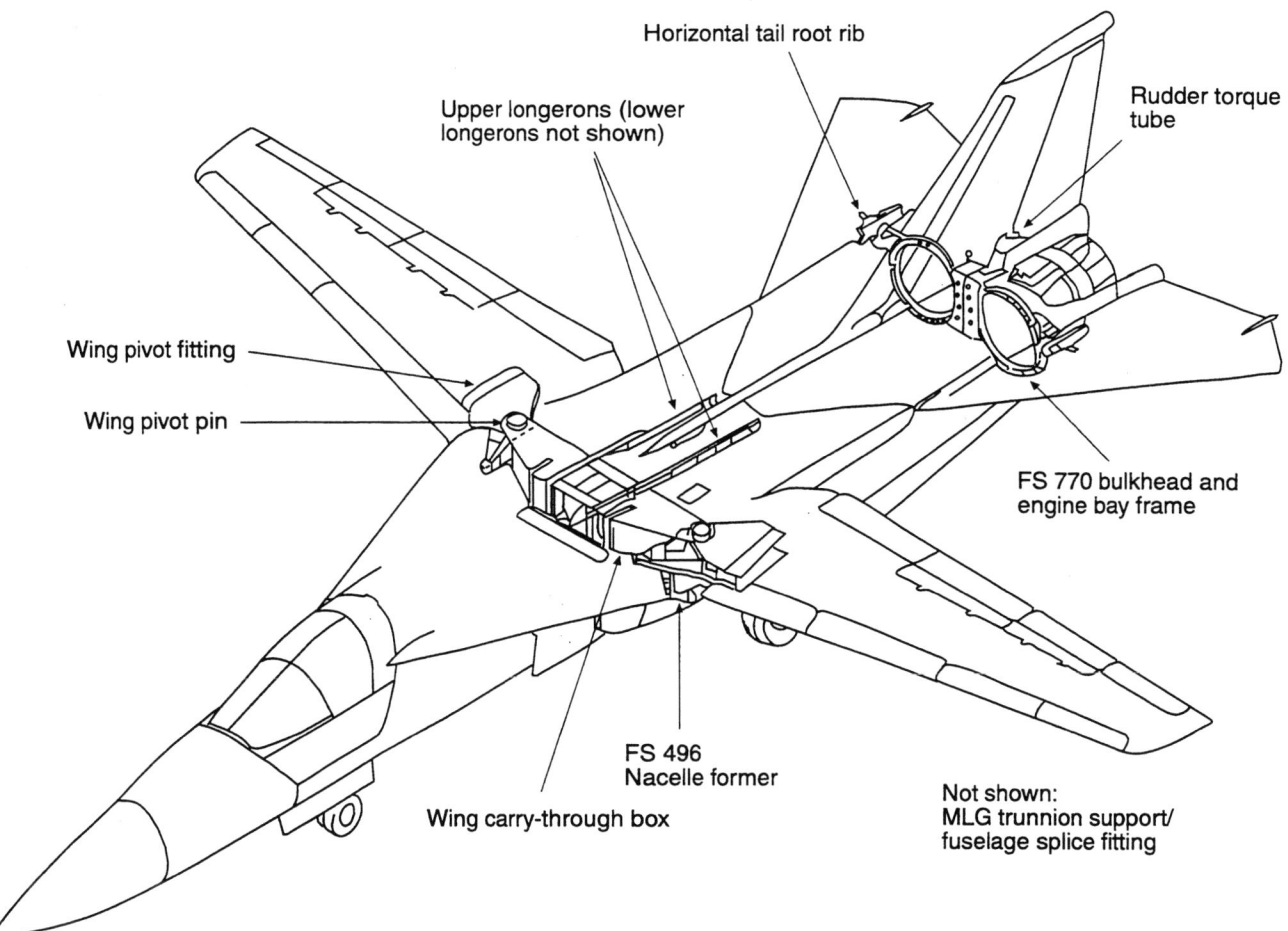

FIGURE A-9 F-111 D6ac steel components.

was completed successfully. A full-scale fatigue test program was initiated in 1968, with the entire program lasting about six years. The program started using a complete airframe test article with the objective of completing four lifetimes of testing or 16,000 test hours of a relatively severe block-type spectrum. However, after only 400 cyclic test hours, the wing carry-through box failed due to a manufacturing quality problem. As a result of the severe damage to the test article, the test program was revised to continue testing on separate major components (i.e., wing, fuselage, etc.) rather than the complete airframe. Also, the test spectrum was revised and the test goal increased to 24,000 test hours. Subsequent failures of the wing carry-through boxes occurred at 2,800 and 7,800 test hours. As a result of these failures, taper-lock fasteners were installed in the lower part of the box and some other design changes were made. At 12,400 test hours, a failure occurred at the wing pivot fitting. This failure resulted in the development of a boron-reinforced composite doubler modification that was the first use of advanced composites to reduce the stress levels in metallic aircraft structures (boron reinforcements were also used later on fuselage longerons on the B-1 bomber). The wing then completed the 24,000 test hours without any further significant events. The test was then continued beyond the contractual requirements to 40,000 test hours. At this point it was subjected to constant amplitude cycling to the bending stress associated with a 5.8-g maneuver load. Failure occurred in the aluminum skin splice to the pivot fitting at 10,151 cycles. The horizontal stabilizers were also subjected to 40,000 test hours and were found to be free of fatigue cracking in a subsequent tear-down inspection. The fuselage/vertical stabilizer article successfully completed the contractually required 24,000 test hours, but failed in an upper steel longeron at 28,800 test hours as a result of fatigue cracking from a fastener hole. This area was modified and the test continued to the 40,000 test hours.

Based on the conventional safe-life approach, the 40,000 test hours should have provided for 10,000 hours of safe operational use using the scatter factor of four. However, in December 1969 a F-111 experienced a catastrophic wing failure during a pull-up from a rocket firing at Nellis AFB.

This aircraft only had about 100 hours of flight time at the time of failure. The failure originated from a fatigue crack that emanated from a sharp-edged forging defect in the wing pivot fitting. This failure graphically highlighted the fundamental shortcomings of the safe-life approach (e.g., it would have required a scatter factor of 400 rather than 4 to prevent this aircraft loss). It was this failure that provided much of the impetus for the Air Force to abandon the safe-life approach and adopt damage tolerance requirements on all of their aircraft in the early 1970s.

As a result of the Nellis accident, the Air Force convened a special ad hoc committee of the SAB to investigate the failure and recommend a recovery program. This committee met with General Dynamics and the Air Force Systems Program Office frequently over a period of 18 months in 1970 to 1971. Early on it was apparent to the committee that it would be very difficult to protect the structural safety of the F-111 using conventional NDE methods because of the low fracture toughness of the D6ac steel and the resulting very small critical flaw sizes and the even smaller flaw sizes that must be found to avoid more failures. This difficulty led the committee to recommend to the Air Force that every aircraft be subjected to a fracture-mechanics-based low-temperature proof load test, which would be repeated at periodic intervals to be determined from the predicted rate of crack growth in the individual aircraft based on its actual measured use obtained from the IATP. This fracture-mechanics-based proof testing concept had been developed and successfully used for the pressurized structure in the Apollo space program as well as in other missile and space efforts. The essence of the concept is that, if the structure successfully survives the proof test load, it could not have contained any flaws larger than the critical sizes at that load level. It is then assumed that it did contain flaws just smaller than the critical sizes at the proof test load level, and to cause failure in service they would have to grow to the critical sizes at the lower operational load levels. The time for this to happen is then calculated for the anticipated use spectrum using crack growth analysis procedures, and the repeat proof test interval is established based on this growth interval. To screen the smallest possible flaw size, the F-111 proof tests were conducted at low temperature (i.e., -40°F), where the fracture toughness of the D6ac steel was lower than at room temperature. In effect, the proof test is a destructive inspection procedure that culls out any flaws that would cause an in-service failure.

Obviously, the desire of both General Dynamics and the Air Force was to minimize the possibility of proof test failure and the associated expenses and downtime. To achieve this objective, they developed a magnetic rubber inspection procedure for detecting very small surface cracks in the steel parts. It has been reported that the process is capable of detecting surface flaws of 0.02 in. long with high confidence. This inspection was conducted on the accessible critical areas of the steel structures at their required inspection interval and prior to each proof test. In total, over 50 specific areas of the aircraft are inspected with magnetic rubber at each depot cycle. Over the years, fatigue cracks have been detected and were subsequently repaired prior to proof testing in 25 different areas of the wing carry-through structure, the wing pivot fitting, the horizontal stabilizer support structure, the fuselage station 496 nacelle former, the main landing gear support structure, and in the fuselage structure. Nevertheless, proof test failures have not been totally avoided. During the past 25 years, there have been 11 proof test failures. There have been two EF-111As, four F-111As, two F-111Es, two F-111Cs, and one FB-111A that failed during proof testing, thus avoiding probable in-service aircraft losses. In fact, since the original catastrophic failure at Nellis AFB in 1969, there have been no F-111 aircraft lost due to structural failure.

During the late 1970s a complete DADTA was conducted on the F-111, and the IATP was upgraded to a program based on crack growth or fracture mechanics, although the proof test intervals had been based on crack growth from the inception of the proof test program. The DADTA initially considered over 400 potentially critical areas, which were subsequently scaled down to about 100 to be analyzed in detail. Currently, approximately 20 areas of the structure are tracked and analyzed and result in periodic updates to the force structural maintenance plan and adjustments in the inspection requirements to account for use changes and base reassignments. Although the first repeat proof testing of the F-111A/E/D aircraft was set at 1,500 hours, this interval was increased to 3,000 hours for subsequent proof tests based on the DADTA and force tracking data. Currently, all of the EF-111As have been proof tested at least three times and some four times.

The F-111 history truly represents a major success story for the Air Force structural integrity program. This success is not only the result of the NDE program, the cold proof testing, the DADTA, and the IATP, but also the capability and dedication of the engineers, managers, logisticians, technicians, and inspectors at the Sacramento ALC and the field-level maintenance personnel. Based on the failure that occurred at Nellis AFB, the many failures encountered during the full-scale testing of the aircraft and the extreme sensitivity of the non-fail-safe structure to small flaws, there was considerable skepticism in the early 1970s with regard to the future structural performance of the F-111 force. Operating and maintaining this force for nearly three decades without further losses due to structural failure has been a significant achievement.

U-2

The U-2 first flew in August 1955 and was introduced in operation in 1956. It was in service for about ten years when the Air Force contracted with Lockheed for an improved

U-2R in 1966. The first flight of the U-2R was in August 1967 and production deliveries started in 1968. In 1979 the U-2 line was reopened, and in 1981 deliveries of U-2 and TR-1 aircraft for the Air Force were started. The TR-1 differed from the U-2 only in some of the on-board systems, and in December 1991 the Air Force decided to redesignate the TR-1 back to the U-2. Currently, the Air Force has 35 U-2s in their active inventory with an average age of 13.9 years; however, they still have four aircraft older than 25 years.

Lockheed-Martin provides logistics support for the U-2 aircraft rather than the Air Force. The committee did not receive a briefing on the structural status of the aircraft from Lockheed-Martin representatives. However, the aircraft was recently reviewed by the Air Force (ASC/EN) and they provided the committee with a summary of their findings. The comments in the following paragraphs reflect these findings.

The U-2 was originally designed (for the intelligence community) and operated before the formation of ASIP and did not comply with all of the ASIP tasks. Also, there has never been a damage tolerance assessment of the airframe either as part of the original design or subsequently. There was no full-scale durability (i.e., fatigue) testing and there was no full-scale static test of the aircraft.

The recent review of the maintenance activity on the U-2 has not shown any evidence of fatigue cracking in safety-critical structure and very little in other structure. Also, there is no evidence of any serious corrosion problems in primary structure. The contractor has found a few isolated areas of corrosion in some areas that were spot welded during the original production of the aircraft. In general, the maintenance of the aircraft has been good and the workmanship of high quality. The wing is constructed of integrally stiffened aluminum with fuel transfer holes in the risers (similar to the C-141). With prolonged use it might be expected that fatigue cracking may initiate at the fuel transfer holes and at the riser runouts such as that experienced on similar designs in other aircraft. There has been some indications of distress in the rear fuselage likely due to hard landings.

The Air Force believes that there is a need for a more-detailed stress analysis of the airframe combined with a flight test program to help identify critical areas and assist in the development of flight-by-flight stress spectra for these areas. They also believe that gust and ground loads will be the major contributing factors in any future fatigue cracking in the airframe. With a knowledge of the critical areas and the stress spectra, the contractor could then make the crack growth calculations, necessary to develop safety inspections and to provide input to a service life estimate for the aircraft. This total effort would then constitute a DADTA comparable to that performed on the other major aircraft systems in the Air Force inventory.

Although the Air Force sees no evidence to indicate that the U-2 aircraft is approaching its life limit, the existing database necessary to make any rational estimate of its future longevity is lacking. Considering the Air Force's desire to retain this aircraft in the inventory for another 15 to 25 years, as indicated in Table 3-1 of this report, the committee believes that it is prudent to perform a DADTA now (as suggested by ASC/EN) before cracking eliminates the opportunities to make economical life extensions and safety modifications (e.g., hole cold working).

AIR EDUCATION AND TRAINING COMMAND AIRCRAFT

T-37

The Cessna Aircraft Company was awarded the contract for two XT-37 prototypes in 1953 and the first flight was in October 1954. The first 11 production T-37s were ordered in 1954 and IOC was achieved in 1957. A total of 534 T-37As were produced. The T-37Bs, which had higher thrust engines, were first introduced in 1959. When production ended in 1977, a combined total of 1,272 T-37s had been produced for the Air Force and foreign countries. As of September 1996 the Air Force still had a total of 420 T-37s in their active inventory. The average age of these T-37s is 34 years. They are scheduled to be retired over the next 12 years and replaced by the JPATS.

The T-37 is essentially a non-fail-safe static-designed aircraft. Fatigue was not recognized as a serious concern by the Air Force until the fatigue failure of the front spar of the wing of an operational aircraft in 1968. This resulted in a modification to the aircraft and the subsequent fatigue testing of two full-scale wing and carry-through structures during 1968 to 1969. The remainder of the airframe (i.e., the fuselage and empennage) was fatigue tested in 1972.

As a result of the full-scale fatigue tests, subsequent fatigue analyses, and analytical condition inspection of operational aircraft, there have been five different fatigue modifications made to the T-37B. One major modification to the wing and forward carry-through structure and the canopy rails was completed in June 1981. This was performed on aircraft with over 5,000 hours of flying time. Another fatigue cracking problem was at the front spar at wing station 46.1. Testing to determine the need for modification of this area was performed; however, the decision was made to inspect rather than modify the area until the aircraft was replaced by the T-46, which was scheduled later in the 1980s. This inspection was implemented in August 1983. The testing also revealed a potential fatigue cracking problem with the wing attachment lugs. This led to the development of a cold working technique, whereby a high interference-fit bushing was installed in the lugs to retard fatigue crack initiation and growth.

When the T-46 program was canceled in the mid-1980s, it became apparent that a complete DADTA was needed for the

T-37. This was performed by Cessna and was completed in May 1988. Eighteen fatigue-critical areas were identified, and safety limits and inspection requirements were established for each. Because of the very short inspection intervals in some areas, a special inspection/modification program called Pacer Deep was performed. This program involved oversizing and cold working fastener holes in several areas of the structure. The objective of Pacer Deep was to ensure safe operations until a structural life-extension program (SLEP) could be completed. Saberliner Corporation and Southwest Research Institute (SwRI) were contracted to develop the SLEP program in 1989 with the objective of modifying the structure such that the T-37Bs could be flown safely for another 8,000 hours. Saberliner Corporation provided modification kits, and modifications were completed in 1994 by the Air Force on 537 aircraft. Full-scale static, fatigue, and damage tolerance tests were performed on a complete aircraft, which contained the modifications. The fatigue and damage tolerance testing went to 32,000 cyclic test hours.

Also, during the 1990s SwRI has been conducting T-37 use surveys at several different Air Force bases (i.e., Randolph, Reese, Vance, and Columbus AFBs). In all of these surveys, aircraft maneuver and control surface parameters were measured along with selected strain gage measurements. In addition, SwRI developed a computer program that generates random flight-by-flight stress sequences for all of the pre- and post-SLEP fatigue-critical locations. These are then used as input to damage tolerance analyses to establish safety limits and inspection requirements (intervals) for the critical areas. SwRI is presently performing a DADTA update that addresses the fatigue-critical areas identified during the T-37 SLEP and those that were previously identified and not modified by the SLEP. A total of 14 areas are being addressed.

T-38

The Air Force ordered three prototype T-38s from Northrop in December 1956 and the first flight was in April 1959. The first production T-38s were delivered to the 3510th Training Wing at Randolph AFB in March 1961. The production of T-38s ended in 1972 after a total of 1,187 had been built for the Air Force plus some for NASA and for foreign military sales through the Air Force. Currently the Air Force has 451 T-38s in the active inventory with an average age of 30 years. The range in flying hours is from 4,441 to 15,432 with an average of 12,234 as of November 1996. The primary uses for the Air Force's T-38s are in specialized undergraduate pilot training and in introduction to fighter fundamentals (IFF).

Structurally, the T-38 is very similar to the F-5A/B aircraft, which was built primarily for foreign military sales. The airframe is a single-load-path design made from 7000-series aluminum alloys, which was typical of other military aircraft designed in the 1950s and 1960s. For example, the lower wing skin was machined from a single plate of 7075-T6 aluminum, which was 0.42 in. thick at the root trailing edge (i.e., wing station 26; see Figure A-10). In 1970 there was a failure of this lower wing skin on an F-5 stationed at Williams AFB in Chandler, Arizona. It was caused by a fatigue crack, which originated at the trailing-edge radius at wing station 26 and grew to a critical size of about 0.20 in. at the time of failure. As a result of this failure, the lower wing skins on the F-5s were increased in thickness so as to lower the stresses by about 20 percent, but, because of their less severe use in the Air Training Command (ATC), no change was made to the wing skins on the T-38s. However, by the mid-1970s there was increasing evidence of potential structural problems with the T-38 that led the Air Force to initiate a detailed DADTA. Some T-38s had been moved from the relatively mild use of the ATC to the Tactical Air Command's more severe lead-in-fighter (LIF) and dissimilar air combat training (DACT) use. Also, because of the fuel shortages of the early 1970s, the F-4s had been removed as the Thunderbird demonstration team's aircraft and replaced by the more fuel-efficient T-38s. Adding to the concern was the fact that, in 1975, the wing on a T-38 at Holloman AFB was found to be cracked in the same area as the aircraft in the 1970 accident at Williams AFB, but fortunately the crack ran into another fastener hole and was temporarily arrested. Also, there had been two fatigue test failures originating in the same area. During the time of the DADTA, an aircraft was lost at Randolph AFB due to a wing failure that, again, originated in the same area. A short time later, an aircraft assigned to DACT use was lost due a wing failure.

In addition to the serious trailing-edge area already noted, the DADTA identified 16 more potentially critical areas in the wing, 15 in the fuselage, and 3 in the empennage. Also, improved finite element stress analyses were performed on the entire aircraft, and stress spectra were developed for the critical areas based on measured LIF, DACT, and Thunderbird use and the improved analyses. Tear-down inspections were performed on 11 wings with about 4,000 to 6,000 hours of ATC plus 500 to 700 hours of LIF, DACT, or Thunderbird use and on 3 wings that had been only in ATC use. These inspections indicated that the wings that had been exposed to the severe use had from 12 to 90 small fatigue cracks in the high-stressed fastener holes in the aft in-board portion of the wing, whereas those that had only ATC use were essentially crack free. The damage tolerance analyses indicated that the critical crack sizes in the aft in-board areas of the wing were very small (e.g., less than 0.10 in. for low fracture toughness 7075-T6 aluminum), making the task of protecting structural safety by inspection nearly impossible. It was apparent that the long-term solution required replacing the lower wing skin with a thicker lower-stressed skin made from a tougher material. Also, it was recommended that the holes in the higher-

APPENDIX A

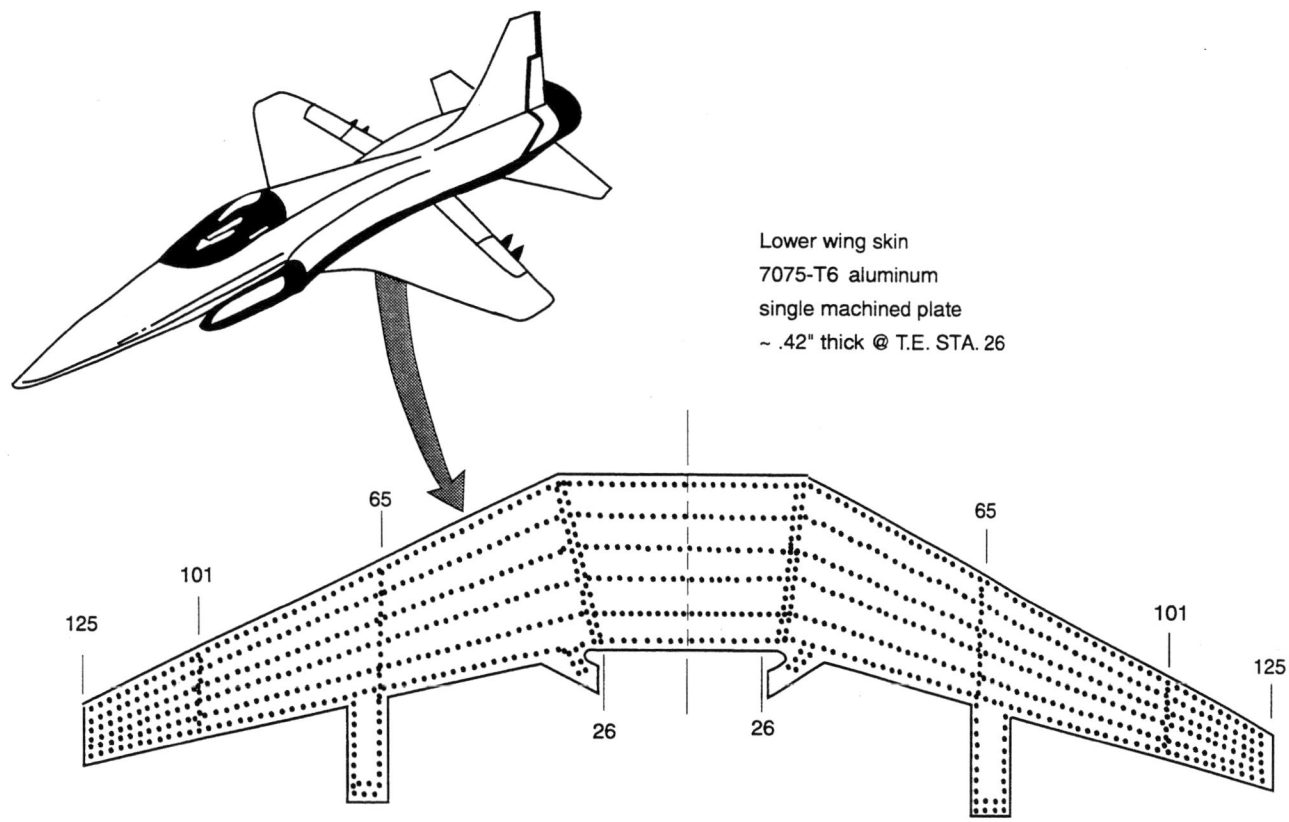

FIGURE A-10 Original lower wing skin design for the T-38 aircraft.

stressed areas of the wing be cold worked. The near-term actions included (1) culling out all wings that had low fracture toughness material using a technique involving measuring the chemical composition of the material and correlating it with a previously established relationship between toughness and chemical composition and (2) using a trained team of NDE specialists to inspect the specific critical locations of the remaining wings using an ultrasonic technique at very frequent intervals.

Since the DADTA and the resulting recovery program in the late 1970s, the lower wing skins have been replaced with thicker skins made from 7075-T73 aluminum, the fastener holes and drain holes in high-stressed areas have been cold worked, and additional full-scale fatigue testing has been performed. Also, the T-38s have been replaced by F-16s in the Thunderbird demonstration team, and the Air Force no longer uses the T-38 in DACT use and has replaced LIF use with IFF use. However, the IFF use is still apparently quite severe. Based on the briefing the committee received on the T-38 structural status in November 1996, the aircraft continue to have more structural problems, and further design changes and full-scale fatigue testing are planned. Specific fatigue cracking problems were identified in the current wings and fuselage of the T-38:

Lower surface of the wing:
- Wing main landing gear door land radius
  - cracking into main skin not repairable
  - modification of land plus special inspection required
  - future design change

- Lower wing skin fastener holes
  - small critical crack sizes (0.2–0.4 in.)
  - oversizing and cold working required
  - future design change

- Wing skin access panel holes
  - D panel; aileron access panel
  - cracking into main skin not repairable
  - stop drill/special inspection (temporary)
  - boron/epoxy doubler (temporary)
  - forcemate bushings under study by the ALC

- Milled pockets on the lower wing skin
  - cracking in milled radius
  - not repairable
  - composite reinforcement under study by the ALC
  - future design change

Fuselage:
- Upper cockpit longerons
  - hookslot cracking
  - small critical crack sizes
  - low inspection intervals
  - material change/redesign
  - force-wide modification in progress

In addition to the fatigue cracking problems, the San Antonio ALC identified the following SCC problems on the T-38:

Fuselage:
- Cockpit longerons
  - upper and lower longerons
  - 7075-T6 aluminum
  - cracking of the forward splice at fuselage station 284
  - inspect and repair (interim)
  - material change/redesign
  - force-wide modification of upper longerons
  - force-wide inspection/repair of lower longerons

- Fuselage forgings
  - bulkheads at fuselage stations 325 and 362; formers at fuselage stations 332, 487, and 508
  - temporary repairs
  - material/design changes
  - force-wide modification of bulkhead at fuselage station 325
  - inspect and repair remaining changes

The following honeycomb deterioration problems were also noted:

- Horizontal stabilizer
  - core corrosion attributable to water intrusion
  - improved bonding being implemented

- Landing gear strut door
  - core corrosion attributable to water intrusion
  - superplastic formed/diffusion-bonded titanium redesign
  - current preferred spare

Based on the above, it is apparent that the San Antonio ALC will continue to face a major challenge to protect the safety and prolong the service life of the T-38 for another 25+ years as indicated in Chapter 2 of this report. Most of the cracking problems noted are safety-of-flight concerns. As noted in Chapter 5, the committee recommends that this aircraft be given high priority in updating the DADTA and obtaining an improved estimate of economic service life.

Appendix B

Biographical Sketches of Committee Members

Charles F. Tiffany (chair) retired as executive vice president of the Boeing Military Airplane Company where he was responsible for the management of several military aircraft programs. His expertise is airframe and propulsion structural design and structural durability and damage tolerance. Since his retirement Mr. Tiffany has been actively involved in the FAA's aging aircraft program and has served on the Technical Oversight Group on Aging Aircraft and the Air Force's Scientific Advisory Board. He is a member of the National Academy of Engineering.

Satya N. Atluri is Institute Professor and Regents Professor of engineering and director of the Computational Mechanics Center at the Georgia Institute of Technology. His expertise is in fracture mechanics, computational mechanics, and the analysis of metallic and composite structures for aircraft applications. Dr. Atluri's research has included the analysis of aged and repaired structures and corrosion–fatigue interactions. Dr. Atluri is a member of the National Academy of Engineering.

Catherine A. Bigelow is manager of the National Aging Aircraft Research Program at the FAA Technical Center. Prior to her present position, Dr. Bigelow spent 15 years at the NASA Langley Research Center. Her primary expertise is in computational mechanics and fatigue and fracture of metallic and composite structures.

Earl W. Briesch is a consultant with Dayton Aerospace Inc. Mr. Briesch is the retired deputy director for requirements of the US Air Force Materiel Command. He has 35 years of experience in logistics management, program management, engineering, and technology on major Air Force weapons systems. He served in senior management positions at Warner-Robins Air Logistics Center with responsibility for major system modifications and depot-level maintenance programs for the F-15, C-141, and the C-130.

Robert J. Bucci is technical consultant at the ALCOA research laboratory. His expertise includes aerospace aluminum alloys, fracture, fatigue, and corrosion. His research concerns in-service degradation of aluminum alloys, fatigue and fracture processes, aging aircraft, materials selection and characterization, and corrosion–fatigue interactions. Dr. Bucci's recent work has concerned the prediction of mechanical performance of aluminum structures with corrosion and multiple-site damage.

Wendy R. Cieslak is manager, Materials Aging and Reliability:Interfaces at Sandia National Laboratories. She is a materials engineer with expertise in corrosion and electrochemistry. Dr. Cieslak's interests include fundamental corrosion processes and corrosion chemistry and her research has concerned electrochemical corrosion, passivation, and advanced battery development.

Eugene E. Covert is T. Wilson Professor of Aeronautics, emeritus, at the Massachusetts Institute of Technology. Dr. Covert's research interests are in aircraft design and aerodynamics and he has broad expertise in the design and operation of military aircraft. He served as chief scientist of the Air Force and chair of the recent Air Force Scientific Advisory Board study, *Life Extension and Mission Enhancement for Air Force Aircraft*, that was a predecessor of this current study. He is a member of the National Academy of Engineering.

B. Boro Djordjevic is associate director of the Center for Nondestructive Evaluation at Johns Hopkins University and president of Materials and Sensors Technologies, Inc. Prior to holding his current position he was manager of Evaluation and Subsystem Engineering at Martin Marietta Laboratories. His expertise is in materials science and engineering, nondestructive evaluation (ultrasonics, acoustics, optical testing), in situ and smart sensors, and advanced composite materials and structures.

Charles E. Harris is chief engineer, Materials Division at the NASA Langley Research Center. He is also the technical manager of the NASA Aging Aircraft Research Program with research in materials characterization and the mechanics of damage in metal and composite structures. Prior to joining NASA, Dr. Harris was professor of aerospace engineering at Texas A&M University and a structural engineer at Babcock and Wilcox Company. Dr. Harris has broad experience in materials and structures, aircraft structural design, and aging aircraft research.

James W. Mar is professor, emeritus, in the Department of Aeronautics and Astronautics at the Massachusetts Institute of Technology. He retired as the Jerome C. Hunsaker Professor of Aerospace Education. His expertise is in the design of aerospace systems such as airplanes and spacecraft, with special interests in structures, aeroelasticity, and materials. He has been a leader in research in aerospace structures and in the use of composite materials in aircraft structure. He is currently chair of the FAA's Technical Oversight Group on Aging Aircraft and has served as the Chief Scientist of the Air Force. Dr. Mar is a member of the National Academy of Engineering.

J. Arthur Marceau is senior principal engineer in Boeing Commercial Airplane Group's Materials Technology organization. His experience has been in corrosion of commercial aircraft structures, corrosion-resistant finishes, and structural adhesive bonding. Mr. Marceau was responsible for the inspection and characterization of corrosion in aging commercial aircraft (under Boeing's Aging Fleet Survey Program) and for the development of corrosion prevention and control procedures for the existing fleet as well as design improvements to avoid corrosion in new aircraft design.

Charles Saff is corporate fellow–concept design and development in the Boeing Information, Space, and Defense Systems Group and is responsible for the development and transition of materials and structures technologies into product applications. His expertise includes materials selection, component design, and analysis of metallic, metal-matrix composite, and polymer-matrix composite structures. Mr. Saff has been involved in materials selection and optimization; integrated design models; and the development of methods for structural risk assessment, failure analysis, and repair of structures for several aircraft programs.

Edgar A. Starke, Jr., is Earnest Oglesby Professor of materials science and engineering and University Professor at the University of Virginia. His research concerns the development and characterization of advanced alloys. He has vast experience in aerospace application of aluminum and titanium alloys and in the characterization of fatigue and fracture processes. Dr. Starke has served on NASA's Aeronautical Advisory Committee, chaired NASA's High Speed Research Committee on Materials, and is currently a member of the NRC's National Materials Advisory Board. He is actively involved in the coordination of international research in aging aircraft through NATO's Advisory Group on Aerospace Research and Development (AGARD).

Donald O. Thompson is director, emeritus, of the Center for Nondestructive Evaluation at Iowa State University. His research interests include materials degradation behavior and quantitative nondestructive evaluation methods and equipment. He has been involved with the FAA's Center for Aviation Systems Reliability at Iowa State University. He is a member of the National Academy of Engineering.